For William

I am diligently forming a library.... The reason for so much labor is to acquire a serene disregard of bad fortune. Not nature alone, as appears to many, commands us to this disregard, but a carefully elaborated system of philosophy.

GERBERT OF AURILLAC, 985

THE **ABACUS** AND THE **CROSS**

The Story of the Pope Who
Brought the Light of Science
to the Dark Ages

NANCY MARIE BROWN

BASIC BOOKS
A Member of the Perseus Books Group
New York

Copyright © 2010 by Nancy Marie Brown

Published by Basic Books,
A Member of the Perseus Books Group

Books published by Basic Books are available at special discounts for bulk
purchases in the United States by corporations, institutions, and other
organizations. For more information, please contact the Special Markets
Department at the Perseus Books Group, 2300 Chestnut Street,
Suite 200, Philadelphia, PA 19103, or call (800) 810-4145, ext. 5000,
or e-mail special.markets@perseusbooks.com.

Designed by Brent Wilcox

Library of Congress Cataloging-in-Publication Data
Brown, Nancy Marie.
   The abacus and the cross : the story of the Pope who brought the
light of science to the Dark Ages / Nancy Marie Brown.
      p.   cm.
   Includes bibliographical references (p.) and index.
   ISBN 978-0-465-00950-3 (alk. paper)
   1. Sylvester II, Pope, ca. 945-1003.   2. Religion and science—
History—To 1500.   I. Title.
   BX1158B76 2010
   270.3092—dc22
   [B]
                                                           2010036361

10 9 8 7 6 5 4 3 2 1

Hildesheim

Magdeburg

Gandersheim

Aachen

ge

Koln

Quedlinburg

RRAINE

Frankfurt

chternacht

Trier

Mainz

dun

Metz

Gorze

Strasburg

Augsberg

Sasbach

Reichenau

GUNDY

Pavia

Bobbio

Piacenza

Ravenna

Rome

SEA

SICILY

Gniezno

POLAND

Prague

HOLY

ROMAN EMPIRE

Gran

HUNGARY

CALABRIA

Rossano

Stilo

N

= Border of the Holy Roman Empire

CONTENTS

Introduction: The Dark Ages   1

PART ONE **From Shepherd Boy to Schoolmaster**
   I   *A Monk of Aurillac*   11
   II   *Of the Making of Books There Is No End*   25
   III *The Ornament of the World*   41
   IV   *The Schoolmaster of Reims*   63

PART TWO **Gerbert the Scientist**
   V   *The Abacus*   79
   VI   *Math and the Mind of God*   93
   VII *The Celestial Sphere*   113
   VIII *The Astrolabe*   133

PART THREE **From Abbot to Pope**
   IX   *The Abbot of Bobbio*   153
   X   *Treason and Excommunication*   177
   XI   *The Legend of the Last Emperor*   199
   XII *The Pope of the Year 1000*   213
   XIII *The End of the World*   231

      ACKNOWLEDGMENTS   251
      ILLUSTRATION CREDITS   253
      NOTES   255
      BIBLIOGRAPHY   279
      INDEX   289

Europe in the Age of Gerbert of Aurillac (950-1003 A.[D.)

ATLANTIC OCEAN

P[

Chartres

Orlea[

Micy

Tours

Nouaille

FRANCE

Aurillac

Santiago de Compostela

Conq[

Toulouse

SPAIN

Cuxa

Nar[

Eln[

AL-ANDALUS

Gi[

Vic & Ripoll

Barcelona

Cordoba

MEDITERRANEA[

AFRICA

The Dark Ages

In the Year of Our Lord 999, the archbishop of Ravenna sat down to answer a letter. He set a sheet of parchment on his tilted writing table, a scrap, off-square, too small to use in a formal manuscript. He crumbled a cake of oak-gall ink, moistening it until it liquefied. He sharpened his goose-quill pen, and sat, and pondered.

In nine months the world would end. There had been famines, floods, comets, eclipses, earthquakes, volcanic eruptions, wolves in churches, rains of blood—so many signs and wonders that they could not be counted. Gog and Magog, in the form of Vikings, Magyars, Saracens, and Huns, besieged Christendom on all sides. Tears flowed from a holy cross. The Virgin Mary appeared in a stone. The cathedrals at Orleans and Mont Saint-Michel were destroyed by fire.

The archbishop of Ravenna knew his Bible: "And he laid hold on the dragon, the old serpent, which is the devil and Satan, and bound him for a thousand years. And he cast him into the bottomless pit and shut him up and set a seal upon him, that he should no more seduce the nations till the thousand years be ended."

But Saint Augustine warned that to guess the mind of God, to think we could predict when those thousand years would end, was blasphemy. Antichrist would come, the dead would rise, Christ would save the good and damn the evil, the earth would be destroyed by fire: "All those events, we must believe, will come about," Saint Augustine wrote, "but in what way, and in what order they will come, actual experience will then teach us with a finality surpassing anything our human understanding is now capable of attaining."

The archbishop of Ravenna's good friend, Abbot Adso of Montier-en-Der, had taken up the question in about 954. Adso sought to soothe the queen of France, who feared the End of the World. Citing the apostle Paul, who had written, "for that day shall not come, except there first come a falling away," Adso argued that "this time has not yet come, because, though we see the Roman Empire destroyed in great part, nevertheless as long as the kings of the Franks who hold the empire by right shall last," the earth would endure.

But the line of Charlemagne failed in 987. The archbishop himself had recorded a "falling away" of churches: Those of Antioch, Alexandria, Constantinople, and the heart of Spain, not to mention Africa and Asia, no longer recognized Rome's oversight.

The archbishop's worst enemy, Abbot Abbo of Fleury, had warned the king of France that rumor of the End Times "filled almost the entire world." In his youth, Abbo wrote, he had heard a priest in Paris claim that the Antichrist would be loosed in the year 1000, heralding the Last Judgment. "I resisted as vigorously as I could to that preaching, citing Revelation and Daniel," Abbo said, but despite his efforts, the rumors would not be suppressed. "Conflict grows in the Church," he warned. A council should be called to decide how to calm the fearful.

"Greed is on the rise and the end of the world is imminent," wrote a scribe.

"Fire from heaven throughout the kingdom, demons appearing," noted an annalist.

"Satan will soon be unleashed because the thousand years have been completed," predicted a chronicler.

"Clear signs announce the end of the world," others concluded: "The ruins multiply."

It was the darkest year of the Dark Ages. Yet the End of the World was not oppressing this archbishop's mind. He was driven to his writing desk by a very different obsession. To his friend Adalbold, he wrote:

> You have requested that if I have any geometrical figures of which you have not heard, I should send them to you, and I would, indeed, but I am so oppressed by the scarcity of time and by the immediateness of secular affairs that I am scarcely able to write anything to you. However, lest I continue mentally disobedient, let me write to you what error respecting the mother of all figures has possessed me until now.
>
> In these geometrical figures which you have already received from us, there was a certain equilateral triangle, whose side was 30 feet, height 26, and according to the product of the side and the height the area is 390. If, according to the arithmetical rule, you measure this same triangle without consideration of the height, namely, so that one side is multiplied by the other and the number of one side is added to this multiplication, and from this sum one-half is taken, the area will be 465. . . . Thus, in a triangle of one size only, there are different areas, a thing which is impossible.

On the eve of the Apocalypse, the archbishop of Ravenna and his friend are discussing the best method for finding the area of a triangle.

It is the last letter we have from this archbishop, Gerbert of Aurillac, before he became pope in April 999 under the name Sylvester II. Before Gerbert's death in 1003, Adalbold would pester the busy man again, this time concerning the volume of a sphere.

For Sylvester II was "The Scientist Pope." To tell the story of his life is to rewrite the history of the Middle Ages. In his day, the earth was not flat. People were not terrified that the world would end at the stroke of midnight on December 31, 999. Christians did not believe Muslims and Jews were the devil's spawn. The Church was not anti-science—just the reverse. Mathematics ranked among the highest forms of worship, for God had created the world, as scripture said, according to number, measure, and weight. To study science was to approach the mind of God.

Gerbert of Aurillac, Pope Sylvester II, left us over two hundred letters and a handful of scientific treatises. He is mentioned in the letters or chronicles of several men who lived during his lifetime. They make it clear he rose from humble beginnings to the highest office in the Christian Church "on account of his scientific knowledge"—not in spite of it. They call him a man of "great genius and admirable eloquence," possessing "incomparable scientific knowledge." He "surpassed his contemporaries in his knowledge," was "acutely intelligent," and "deeply learned in the study of the liberal arts." He was the leading mathematician and astronomer of his day.

From their writings and his own, Gerbert's biography has been known to historians for hundreds of years. Some overlooked it. Some twisted it to their own ends. Others suppressed it—for the picture Gerbert paints of the Dark Ages is lovely and surprising. His world was one in which our modern tensions—Christianity versus Islam, religion versus science—did not yet exist.

Born in the mountainous Cantal region of France in the mid-900s, Gerbert entered a monastery—the only elementary school of his day—to learn to read and write in Latin. He studied Cicero, Virgil, and other classics. He impressed his teacher with his skill in debating. He was a fine writer, too, with a sophisticated style graced with rhetorical flourishes.

To further his education, his abbot sent him south to the border of Islamic Spain, then an extraordinarily tolerant culture in which learning was prized. In the library of the caliph of Cordoba were at least 40,000 books (some said as many as 400,000); Gerbert's French monastery owned fewer than 400. Many of the caliph's books came from Baghdad, known for its House of Wisdom, where for two centuries works of mathematics, astronomy, physics, and medicine had been translated from Greek, Persian, and Hindu and further developed by Islamic scholars under their caliph's patronage. During Gerbert's lifetime, the first of these science books were being translated from Arabic into Latin through the combined efforts of Muslim, Jewish, and Christian scholars. Many of those interested in the new sciences were churchmen, and some became Gerbert's lifelong friends and correspondents.

A professor at a cathedral school for most of his career, Gerbert was the first Christian known to teach math using the nine Arabic numerals and zero. He devised an abacus, or counting board, that mimics the algorithms we use today for adding, subtracting, multiplying, and dividing—it has been called the first counting device in Europe to function digitally, even the first computer; in a chronology of computer history, Gerbert's abacus is one of only four innovations mentioned between 3000 B.C. and the invention of the slide rule in 1622.

Like a modern scientist, Gerbert questioned authority. He experimented. To learn which rule best calculated the area of the equilateral triangle that he had sent to Adalbold, he cut out identical little squares of parchment and measured the triangle with them. To learn why organ pipes do not behave acoustically like the strings of a lyre or harp, he built models and devised an equation. He made sighting tubes to observe the stars and constructed globes on which their positions were recorded relative to lines of celestial longitude and latitude. He (or more likely his best student) wrote a book on the astrolabe, an instrument for telling time and making measurements by the sun or stars—you could even use it to calculate the circumference of the earth, which Pope Sylvester and his peers knew very well was not flat like a disc but round as an apple. Gerbert made an armillary sphere—a primitive planetarium—to explore how the planets circled the globe of the earth; he even knew Mercury and Venus orbited the sun.

For his royal patrons he built siege weapons and pipe organs, dabbled in poetry and astrology, and organized scholarly debates. But most of all, as a sought-after teacher, he spread the science of Islamic Spain throughout Christian Europe. He taught future abbots, archbishops, kings, popes, and emperors.

Brilliant, curious, systematic, and high-minded, Gerbert was less successful in politics. Though he climbed to spectacular heights—

abbot, archbishop, tutor and counselor to emperors and kings, even pope—his progress was erratic. Twice he was accused of treason, each time to be rescued by the sudden, suspicious death of his king. Twice he was forced to flee for his life, once under sentence of excommunication.

From Spain he had gone first to Rome, where he impressed the pope and Emperor Otto the Great with his learning. He was assigned, briefly, to tutor the emperor's son, Otto II. Ten years later, happy as a schoolmaster at the famous cathedral of Reims near Paris, Gerbert came again to Otto II's attention. Now emperor, Otto II appointed him abbot of the monastery in Bobbio, Italy. Bobbio had the best collection of books in Christendom, but politically it was a snake pit. When Otto II died three years later, Gerbert abandoned Bobbio and fled back to Reims.

He longed to resume his scientific studies; instead he became enmeshed in intrigues. He wrote persuasive letters and worked as a spy. His machinations gave imperial rule to the three-year-old Otto III and his Byzantine mother, Theophanu, in place of their bellicose challenger, Henry the Quarreler. His efforts ended the dynasty of Charlemagne, raising Hugh Capet to the French throne. A grateful King Hugh made Gerbert archbishop of Reims, when the post became vacant, but the pope refused to acknowledge him. Pope and king fought over him for seven years. Excommunicated by the pope, Gerbert was abandoned by King Hugh's son and successor. He fled again, this time to Otto III's court, where he dazzled the teenaged emperor with his scientific brilliance.

Otto took Gerbert on as his teacher, then as his friend and counselor. Gerbert's excommunication was reversed by a new pope, Otto's cousin, who made him archbishop of Ravenna. When that pope suddenly died, Otto III advanced Gerbert to the papacy itself. On April 9, 999, Otto's army saw Gerbert installed as Pope Sylvester II.

The two, emperor and pope, shared a dream. Gerbert encouraged Otto to see himself as a second Charlemagne—one with royal Byzantine blood. Otto could reunite Rome and Constantinople, expanding the Holy Roman Empire (then just parts of Germany and Italy) to recreate the vast unified realm of the Caesars. Otto and Gerbert brought two of the scourges of Europe—the Vikings in the north and the Hungarian Magyars in the east—into the Christian fold. They established the Polish Catholic Church and sent missionaries to the Prussians, Swedes, and other pagan tribes; they strengthened the empire's ties with Spain and made overtures to Constantinople. But Otto died in 1002, just twenty-two—and Gerbert a year later, some say of grief.

Their plans for a Christian empire based on peace, tolerance, law, and the love of learning died with them. The Great Schism of 1054 permanently divided the Roman Catholic and Eastern Orthodox churches, and the First Crusade in 1096 redefined the relationship between the Christian and Muslim worlds. Just before the crusade, Gerbert was branded a sorcerer and devil-worshipper for having taught the mathematics and science that had come to Christian Europe from Islamic Spain. Instead of lovingly collecting, copying, and translating the wisdom of Islam, the monks of Christendom began mutilating scientific manuscripts, erasing pages of what they now considered useless information and writing over them. The interests of the Church had changed. Science had lost its central place. Much of what Pope Sylvester knew would be forgotten for hundreds of years.

But Gerbert's teaching, and the books written by his pupils and peers, enabled scholars during the Renaissance to rediscover the math and science he knew so well. Given his tarnished reputation, they did not think to credit him or his sources. Consequently, most people have no idea that our modern technological civilization depends on the science of Baghdad's House of Wisdom, brought to Spain by Muslim scholars and spread through the West—by visionary Christians such as Gerbert of Aurillac—before the year 1000. To tell his story is to look back a thousand years and see an opportunity that was missed. During Gerbert's lifetime, science transcended faith and faith encompassed science: The pope studied the stars and found God in numbers.

PART ONE

FROM SHEPHERD BOY TO SCHOOLMASTER

Equally in leisure and in work we both teach what we know, and learn what we do not know.

GERBERT OF AURILLAC, 985
(quoting Cicero)

CHAPTER I

A Monk of Aurillac

The castle of Aurillac crowned the hill above the river Jordanne, keeping watch over the monastery at its feet. From his keep Count Gerald the Good could look north to the jagged Cantal peaks, snowcapped as late as May. South of the mountains, themselves the southern edge of the Massif Central of France, his holdings stretched a hundred miles toward the Mediterranean, from high mountain plateaus through rumpled hills and river gorges to steep, secret valleys. He owned villas and churches, vineyards and forests, pastures and quarries. Many of his estates were worked by slaves, though here and there a parcel was farmed by free peasants, heirs of Roman colonists. All were interspersed, patchwork style, with the estates of other knights and castellans—his enemies.

Gerald did not want to be a count. He wanted to be a monk. But as his noble father's only child, he was compelled, says his medieval biographer, "to be occupied in administering and watching over things." It was "more holy and honest," he was told, "that he should recognize the right of armed force, that he should unsheathe the sword against his enemies, that he should restrain the boldness of the violent." He could not be a monk.

But he could found a monastery.

Tales of Count Gerald, the spiritual, monkish knight who one day would be a saint, were among the first stories, outside the Bible, that Gerbert of Aurillac heard. Growing up in Saint-Gerald's monastery,

11

young Gerbert learned to see the good count as a hero, a role model for a man of God in a lawless age.

By Count Gerald's day, in the late 800s, the king's justice was a fond memory. Five hundred years after the fall of the Roman Empire, the old Roman law was still revered in Aurillac, a town named for Marcus Aurelius. Count Gerald could cite the laws of Caesar Augustus—but he could not enforce them outside his own county. The king was far off and feeble. The grandsons of Charlemagne had split up his empire, and their successors were weak-willed and short-lived.

Viscounts promoted themselves to count, counts to duke, and anyone who could afford it built himself a castle and called himself a castellan. No king curbed their ambitions or their feuds. The rampaging troops of Duke William I of Aquitaine (called "the Pious") looted and lay waste to whole regions. Knights fired fields, rustled livestock, and ransacked churches; a priest who objected had his eyes put out; a peasant who refused a trumped-up tax saw the same done to his young son. A neighboring castellan issued from his motte-and-bailey fortress like a wolf in the evening to attack passersby. Robbers haunted the woods.

To right such wrongs, says his medieval biographer, Count Gerald the Good was "kept in the world." He chose nonetheless to live like a monk. He learned his letters and Latin grammar. He chanted the psalms. He ate little and never drank to excess. He allowed no chattering or buffoonery at table or (God forbid) music of harp or lyre, but took pleasure in conversations on practical or spiritual topics and in the reading aloud of books. Three days a week he ate no meat. Throughout his life he remained chaste; he never married. Once he had built the monastery at the base of the castle hill, the count ornamented its church with saints' relics, and bequeathed it most of his property. He went often to Rome and gave generous alms along the way. He fed and clothed the poor. He himself wore only wool or linen—no silk—and as for jewels, but one gold cross.

Miracles happened. When he was forced to fight, "Count Gerald the Good commanded his men in imperious tones to fight with the backs of their swords and with their spears reversed. This would have been ridiculous," admits his biographer, if he had not been invincible.

On one occasion a neighboring count attacked the castle of Aurillac in Gerald's absence and stole everything he could carry away. The next time he was not so lucky: Gerald was in his chapel. His knights, hearing the outcry, begged to go fight, but Gerald insisted they finish Mass. The gates, providentially, were shut, and the attackers, finding nothing else to steal, took seven horses. Soon after, sixty of the attackers' own horses died. Terrified, they returned the seven from Aurillac.

Then there were the blind and the lame, said to be healed by the water in which Gerald had washed his hands.

Count Gerald—Saint Gerald, as he would become—was the model of a perfect, gentle knight. *The Life of Saint Gerald of Aurillac*, written by the influential Abbot Odo of Cluny shortly after Gerald's death, became the foundation for the medieval code of chivalry. To it can be traced Gerbert's deepest values and yearnings: for holiness allied with strength, and law backed up by learning.

Gerbert was born in Gerald's domain about forty years after the count's death in 909—no one knows exactly when or where. Folklore places him in the hamlet of Belliac, a cluster of low stone cottages beside a spring, the castle of Aurillac within sight to the south, a motte-and-bailey fortress around a bend to the north. Since the 1400s, a house in Belliac has been called "the house of the pope." A cock crowed three times when Gerbert was born there, the story goes, and the noise carried all the way to Rome.

Another tale names Gerbert a *pastor*'s son: In medieval Latin, *pastor* often refers to an abbot or bishop. Celibacy was not required of the higher clergy in Gerbert's day: They took no such vows. Though they could not marry—for economic reasons—they had "housekeepers." Their children, who could not inherit, were often given to a church, as were the bastard children of kings and nobles. The father need only draw up a contract and, before witnesses, wrap it and the infant's hand in the church's altar-cloth: The child was committed, for life.

But *pastor* literally means "shepherd." Gerbert could have been the son of a shepherd, a free peasant who had inherited his land and worked it without any obligations except his tithe to his church and taxes to the local lord. Free peasants were still numerous in this part of France in the mid-tenth century. Society there remained divided into two classes: free or not free. Feudalism, which broke society into three groups—churchmen, nobles, and serfs—was only just beginning to catch on. Not until the next century would the aged Bishop Ascelin of Laon, once a student and sometimes-friend of Gerbert's, codify the new rule: Just as God is three-in-one, here on earth some pray, some fight, and some work. "These three are one, indivisible," the bishop wrote, "for each one supports the work of the other two."

Free peasants disappeared under the new feudal code. Slaves, the unfree, were better off, having gained at least some rights. But peasants lost rights they had long held. True, a peasant with horse and armor—plus the youth and strength and character to fight—could become a knight, and thus a nobleman, under the new feudal system. But a peasant who was too poor or old or meek lost the right to bring a lawsuit to court or bear witness for a neighbor; he lost even his land, bartered to a castellan in exchange for protection by (or from) his knights. As feudalism took hold, the peasant was demoted to a serf. No longer permitted to leave his farm, he was deprived even of the right to marry off his children unless his lord permitted the match. If the land was sold, the serf and his family went with it. This societal shift was already beginning in Count Gerald's lifetime and would continue throughout Gerbert's.

All we know for certain about Gerbert's rank is what he tells us himself. Writing years later, he sounds amazed that he could ever have been elected archbishop of Reims: "I do not know, I repeat, I do not know, why I, destitute and an exile, aided neither by birth nor wealth, was preferred to many persons who were wealthy or conspicuous for the nobility of their parents, unless by Thy gift, good Jesus, who lifts up the poor out of the dunghill to sit with princes."

If he was indeed a peasant boy, Gerbert's luck changed one day when the abbot of Saint-Gerald's monastery stopped by the meadow at Belliac, says another folktale. Chatting with the shepherd lad, the abbot (also named Gerald) was so impressed by his intelligence that he offered Gerbert a place in the monastery school. Or so the story goes. One way or another, Gerbert did enter the monastery. It was the only way he could get an education: The Church ran the only schools.

Gerbert loved his years at Saint-Gerald's. Many monks lamented the tedious days hunched over books, the endless cold of the scriptorium, the teachers who smacked their wrists with a rod, the strain of getting out of bed for nighttime prayers; Gerbert never complained. In the 233 letters he left us, he never once mentions his parents or kin. Yet of Abbot Gerald he writes: "Happy day, happy hour, where I am permitted to know a man the remembrance of whose name alone suffices to make me forget all my pain. If I could only see him oftener I would be the happier."

All monasteries in Gerbert's day were Benedictine, guided by the sixth-century Rule of Saint Benedict. According to the Rule, a monk was to be content with the poorest and worst of everything. Yet to a peasant boy in tenth-century France, sleeping alone in a bed with a pillow, a candle burning all night, was luxurious. He would not have despised a monk's clothes: a light shift with wide sleeves, a woolen tunic, a pair of trousers, a thick wool cowl for winter and a lighter one for summer, shoes and socks, a fur cap, winter boots and mittens, two heavy cloaks (of leather and fur), and a leather belt. Coarse and cheap, they were nevertheless adequate—and when they were outgrown or torn, they were replaced. Nor would he have minded (if he was like a modern boy) having a bath only three times a year.

The food at the monks' table would have delighted him. It was no hardship to eat one hot meal a day in winter and two in the summer, when the days were longer, along with any fruits or vegetables that were in season, a pound of bread, and as much wine as the abbot thought he needed, sometimes spiced or sweetened with honey. Every Sunday, each child got a cup of milk. A common supper was made of five eggs and cheese, cooked and served with a side of fish—for each person. On fast days, beans and vegetables replaced the eggs and cheese, but the fish was still allowed.

Salmon, pike, trout, eel, lamprey, and squid were on the menu, seasoned by mustard seed and vinegar brought around on a tray by a serving-monk. Though monks ate red meat only when they were sick, few peasants could afford it even then. And having the allowance to *be* sick—and be nursed in the monastery's infirmary—must have amazed him.

In terms of behavior, however, the monastery at Aurillac was strict, hewing closely to the Rule—which should better be called the "Rules," for it contains many of them. A monk was not drowsy, not lazy, not a grumbler, but obeyed his superiors without hesitation, delay, or objection. He did not speak until spoken to, going about his business always with head bowed and eyes on the ground. When he did answer a question, he did not laugh or raise his voice, but answered humbly, in a few sensible words, for "a wise man is known by the fewness of his words," the Rule said.

The rule of silence was a signature of the Benedictines. Only during two short periods, morning and evening, was conversation allowed. At all other times, the monks used sign language—what one called "the language of the fingers and the eyes." At meals, waving both hands in a circle, with thumbs and two fingers raised, meant *pass the bread*. For fish, wiggle one hand like a swimming fish. For milk, touch the lips with a little finger. For wine, bend your finger and then touch it to your lips. For honey, lick your fingers. For pancakes, ruffle your hair. There were signs for clothes and bedding, books and blessings. "Pillow" combined the sign for "sleep" (a hand against the jaw) and the one for "Alleluia": "Raise your hand, bend the tips of your fingers, and move them as though for the purpose of flying," says the *lexicon*, or sign book, of the monastery at Cluny, because the angels in heaven sing Alleluia (and, presumably, their feathers stuffed monks' pillows). Even ideas could be conveyed with signs. For "hearing," hold a finger against your ear. For "I don't know," wipe your lips with a raised finger. To signal that someone (else) is telling a lie, "place your finger inside of your lips and then draw it out again." To say something is good, place your thumb on one side of your jaw and your fingers on the other and stroke down. For "bad," spread your fingers over your face and pull them away, fast, like the claws of a bird.

The rule of silence permitted monks something few had at the time: solitude. Privacy was rare in medieval life. Peasant families (and their cows and chickens and pigs) slept and ate and worked in one-room huts. Town-houses were small and tightly packed inside town walls. Knights crowded into halls with their wives, children, servants, kinsmen, neighbors, dogs, and hangers-on. A monk, too, was rarely out of sight of another monk. But in the quiet of the cloister, he could at least be alone with his thoughts.

Other rules might especially tax a quick-thinking boy like Gerbert. In later life, when he was no longer a monk, Gerbert would break all of these: not to be proud or haughty, not to give way to anger, not to be jealous, not to nurse a grudge; to always defer to his elders; and, most of all, to attribute to God and not to himself whatever good he saw in himself.

Infractions incurred first a warning, then a public rebuke. If he still did not reform, the monk could be excommunicated—shunned: No other monk was allowed to eat with him, sit with him, pray with him, work with him, or talk to him, even by signs. "He shall not be blessed by those who pass by, nor shall the food that is given to him be blessed." The sentence was not lifted until the wrong-doer prostrated himself before the door of the chapel, lying face to the ground as the others came out of church, and the abbot was satisfied that his display of humility was sincere (which meant this could go on for days). Those too young or "perverse" to understand the seriousness of excommunication were whipped.

Boys Gerbert's age were also whipped for mistakes they made reciting a psalm, a responsory, an antiphon, or a lesson during one of the seven church services every day. They were whipped if they failed to get up promptly for (and stay awake during) the night offices of nocturnes and lauds, which began in the wee hours. They were whipped if they did not attend prime at sunrise, terce in mid-morning, sext at noon, none in the mid-afternoon, vespers at sunset, and compline at dark, after which they were whipped if they did not go straight to bed until nocturnes came around again. Why were the monks so strict when it came to the daily rites? This continuous conversation with God was the purpose of a monastery.

For medieval thinkers, salvation—heaven—was a matter of economics. Simple repentance was not enough; sin had to be paid for. The sinner had impugned God's honor and must pay compensation. The concept was familiar from codes of law, particularly those of the north. A Viking who stole his neighbor's horse was fined three marks of silver (worth nine yards of homespun cloth), in addition to having to return the horse; if he killed a man, he owed twenty-five marks to the next of kin. A man "who vomits on account of drunkenness," says a seventh-century penitential from Canterbury, was assigned fifteen days of penance, which meant he had to fast on bread, water, and salt and abstain from taking the sacraments at Mass. Sex with a virgin required a year's penance; killing a man, seven years.

The Church had little power to enforce these rules; they rested on the conscience of the sinner. If you were not careful in your accounting, you could pile up several years' worth—a lifetime's worth—of penance. What happened if you died first? The thought haunted medieval men and women. Purgatory was not invented until the twelfth century. Instead, sinners of Gerbert's time were taught that good works—pilgrimage, almsgiving, endowing a monastery, and buying monks' prayers—could replace days of penance. Monks like Gerbert, with their seven daily Masses, were surrogates for the sinners of the world.

Consequently, the cloister was not as cut off from the world as we might suspect. Young Gerbert had chances to see (if not talk to) all sorts of new faces, as Saint-Gerald's competed to attract paying sinners.

The Rule had long obligated monasteries to care for travelers, whether on pilgrimage or not. It released the abbot from his vow of silence so he could suitably entertain his guests. But what began as charity, with no payment expected, by the tenth century was an industry. To lure paying pilgrims, monasteries on the roads to Rome or Jerusalem built guest houses that were first-class hotels, providing rooms, meals, storehouses, and stables for everyone from peasants to princes. The guest house at the famous monastery of Cluny was a palace holding forty-five beds for men and thirty (in a separate wing) for women.

But why go all the way to Jerusalem or Rome when it was just as good for your soul to visit saints' relics—and buy Masses—closer to

The church of Saint-Michael-of-the-Needle, from which Bishop
Godescalc of Le Puy began his pilgrimage to the shrine of Saint
James at Compostela, pioneering the route that passed through
Aurillac during Gerbert's childhood.

home? That was the thought of Bishop Godescalc of Le Puy, who in
951 (about the time Gerbert was born) was the first pilgrim to tread
the road to Compostela in Spain, where the body of Saint James the
Apostle, Santiago, had allegedly been found about a century earlier. Le
Puy was a hundred miles from Aurillac, to the northeast across the Cantal peaks, and Godescalc's path took him from his fortress-like church,
built crowning a rock called "The Needle," through the mountains to
Saint-Gerald's, where the good count's bones were continuing to heal
the blind and lame after his death. Hundreds and thousands of pilgrims, over the centuries, followed the route the bishop had made famous: All stopped at Aurillac. Though today Aurillac is considered the

most out-of-the-way of France's provincial capitals, in Gerbert's day it sat on the central highway.

From Saint-Gerald's, Bishop Godescalc traveled south to Conques, which held the miracle-working relics of Saint Foy. Her story illustrates the stakes involved in monastic competition—and the lengths to which some monks would go. In the fourth century, some 130 miles away in Agen, this thirteen-year-old Christian girl had been martyred for her faith by the Roman governor. Her relics did not arrive in Conques until 866. One story says they were sent there for safekeeping from the Vikings. Another, from *The Book of Miracles of Saint Foy*, says that a monk from Conques came to Agen and, insinuating himself into trust, was appointed sacrist, the guardian of the relics. "One stormy night he stole them."

At Conques, Saint Foy made deaf-mutes speak, cripples walk, and the blind see. Prisoners who prayed to her were freed like Houdini from every fetter their captors could apply. But Saint Foy also had a wrathful side. When her bones were carried past in a grand procession—the monks holding crosses and Bibles and vessels of holy water, clashing cymbals, and blowing on ivory horns—a girl who refused to bow was turned into a cripple. A monk who refused to worship her met the saint in a dream. She beat him with a sturdy rod; he died after relating the tale. Another was struck by lightning.

News of Saint Foy's power spread. Alms rushed in, from finger-rings to manor houses. "Though the abbey had long ago been poor, by these donations it began to grow rich and to be raised up in esteem," *The Book of Miracles* says. Its treasury soon held a huge silver crucifix and too many gold and silver reliquary boxes to count (including one containing the Holy Foreskin), along with basins, chalices, crowns, candelabra, thuribles for burning incense, silver frontals for the minor altars, and a great golden frontal, seven feet long, for the high altar. "Few people are left in this whole region who have a precious ring or brooch or armbands or hairpins, or anything of this kind," reports *The Book of Miracles*, "because Saint Foy, either with a simple entreaty or with bold threats, wrested away these same things. . . . She demanded no less from the pilgrims who pour in from every direction."

Before Saint Foy arrived, Conques had been slated to close. Already in the 700s, the king had found it too far off the beaten track, hidden in its tiny, shell-shaped valley, and had established a new monastery, Figeac, up on the plateau, more convenient to the royal itinerary. The relics put an end to these plans. Figeac had to come up with its own saint just to stay in business. Abbot Haigmar, who "was always eager to acquire the bodies of saints by trickery or theft," according to a tenth-century account, soon acquired Saint Bibanus. He was aided by Viking raiders who had, providentially, just sacked the city of Saintes, where Bibanus had been an early bishop. In the confusion, the Figeac monks opened the saint's grave and spirited off his holy bones.

At Conques, Saint Foy occupied a niche of her own. Her reliquary was the kind known as a "majesty": a full figure of the girl, seated on a throne, face forward, arms out, knees rigid, the whole about two feet tall (see Plate 1). Carved of wood and plated entirely with gold, it was begun in 942 and finished in about 984. As the fame of Saint Foy's miracles increased it became more and more encrusted with precious stones and cameos, many of them recycled from Roman jewelry. Her head was recycled too: It came from the golden statue of a Roman man, the only suggestion of femininity being the dainty earrings in her ears.

This very adult, mannish Foy was so popular with pilgrims that Aurillac was forced to compete. The abbey established its own gold workshop and made a matching majesty of Count Gerald the Good—by then known as Saint Gerald. The workshop may have been active before Gerbert left Aurillac as an adolescent in 967: He shows a grasp of metalworking in his later creations of scientific instruments.

The Book of Miracles praised Saint Gerald's golden likeness in 1010: "It was an image made with such precision to the face of the human form that it seemed to see with its attentive, observant gaze the great many peasants . . . praying before it," says the astonished author, Bernard of Angers. To his companion, Bernard says, "I burst forth in Latin with this opinion: 'Brother, what do you think of this idol? Would Jupiter or Mars consider himself unworthy of such a statue?'" Even as they marveled at its realism, the two monks agreed it was absurd "that so many rational beings should kneel before a mute and insensate thing." It is the kind of response Gerbert might have had—he was always rational, never mystical in his approach to worship. And, like Bernard, he may have felt the tension in the Church. "If I had said anything openly then against Saint Gerald's image, I would probably have been punished as if I had committed a great crime," Bernard concludes. For the cult of the saints was growing in popularity and political importance throughout the Church. Saints were, as Bernard's *Book of Miracles* admitted, good fund-raisers.

The size of its church also advanced a monastery's reputation. The rush of pilgrims that followed in Bishop Godescalc's footsteps along the road to Compostela inspired the abbot of Conques to expand his sanctuary, adding a passageway behind the altar to control the traffic flow to Saint Foy's majesty. The abbot of Aurillac, too, started a church-building program. A stone-carver's workshop joined the cluster of buildings at the base of the castle hill; it would become renowned for the detailed *palmettes*, interlaced knots, and beaded ribbons carved on the capitals of columns. The new church would not be finished until 972—five years after Gerbert left Aurillac—so construction was for him an ordinary part of monastery life; he could have learned much about stonework, engineering, and architecture, just by watching.

The new cathedral was built onto and around the old one. Count Gerald's original church looked like any small mountain church in southern France: a single square nave with a semicircular apse that held the altar. The church of 972 was a basilica, with a much longer nave divided into three aisles, the two side aisles having lower roofs so that light could shine into the center from high clerestory windows. Where it met the apse, the nave opened out to left and right, making the shape of a cross. One arm led to a side chapel, the other to a bell tower. The walls would have been fortress-thick, of rough, unshaped stone and rubble, held together with a generous amount of mortar, like the church of the same age at Cuxa in the Pyrenees. Like the cathedral of Santa-Maria in Cosmedin in Rome, restored to its tenth-century splendor, the church had been painted with brilliant frescoes—the Lamb of God in a blazing blue dome, surrounded by swirls and curlicues of red and green, bright biblical scenes in panels on the walls below. Saint Gerald's remains in their golden majesty were placed on the altar, surrounded by other relics of saints. At Cuxa, the altar was a vast slab of white marble, seven feet long by four and a half feet wide, that had come from Roman ruins. It bore a morsel of the true Cross and eighty-nine other relics. The altar at Aurillac would have been similarly splendid.

Church building did not excuse the monks from their duties. The seven daily services went on as best they could in the old church even as it was being engulfed by the new. In between church services, Gerbert and his fellow monks washed and ate, attended chapter (during which wrong-doers were expected to confess their faults), and read. Two hours a day, at minimum, Gerbert would have spent reading—more on Sundays and during Lent, when each monk was given a book to read "straight through," the Rule says, adding, "If anyone should be so negligent and shiftless that he will not or cannot study or read, let him be given some work to do so that he will not be idle."

Not all the books were sacred texts. A list of 63 books handed out one Lent in the eleventh century at the Italian abbey of Farfa shows the monks read the *Lives of the Saints* and biblical commentaries by Saint Jerome and Gregory the Great. They also read Bede's *History of the English* and Livy's *History of Rome*. At Cluny, they read histories by Bede, Eusebius, Josephus, Livy, and Orosius. The German monastery of Reichenau owned 415 books in the mid-800s. Along with religion and history, there were books on architecture, medicine, and, especially, law, including Roman law, Germanic codes, and the laws of Charlemagne and his successors, the Carolingians. The largest library in the Christian West in the late tenth century was at the abbey of Bobbio, Italy. Among its 690 books were Virgil, Horace, Lucan, Ovid, Juvenal, Martial, Persius, Claudius, Lucresius, the *Cycle of Troy*, the *Legend of Alexander*, Porphyry's *Introduction to the Categories of Aristotle*, Boethius's *On Arithmetic*, a book on diseases of the eyes by Demosthenes, and one on cosmography by Aethicus.

Many of these books were gifts from monks or their families. When Odo entered Cluny in 910 he brought the monastery 100 books. Odo's father was a lawyer who liked to read ancient histories, so some may have been history or law books. Odo also knew the classics, but, according to a famous passage in *The Life of Saint Odo*, he rejected them: "When he wanted to read the songs of Virgil, there was shown him in a vision a certain vessel, most beautiful indeed outside, but full of serpents. . . . From this time onwards he left the songs of the poets, and, taught by the Spirit from on high, he turned his attention wholly to those who expounded the Gospels and the prophets."

Odo may have rejected Virgil—but not before he had read him. Nor did his vision of serpents keep Odo from going to Paris to study the works of Martianus Capella with the renowned scholar Remigius of Auxerre, the *Life* admits. Martianus Capella's *The Marriage of Mercury and Philology*, written in the fifth century, packages a textbook on the seven liberal arts inside a colorful fable. It is full of allusions to classical tales of gods and heroes. Tenth-century manuscripts are likewise rich with glosses and marginal explanations of Greek names—proving their monkish readers did not just skip the pagan foolishness and zero in on the lessons in rhetoric or geometry. According to one gloss, truth needed such fables to give it eloquence, for without eloquence, the naked truth would neither be understood nor believed.

Odo of Cluny, who wrote *The Life of Saint Gerald of Aurillac*, was remembered as a great religious reformer. During his years as abbot, from 926 until his death in 942, he traveled widely, bringing many monasteries—including Aurillac—under Cluny's influence. Everywhere he reinstated a strict observance of the Rule of Saint Benedict. He was also a dedicated reader, known to ride with the reins of his mule in one hand and an open book in the other. In his reforms he did not overlook the Rule's emphasis on reading and study, nor the single technology that made education in the Dark Ages possible: book production. Virgil's serpents notwithstanding, the Latin classics exist today only because reformers, such as Odo, saw to it that Gerbert and his brother Benedictines *did* read them—and make copies.

CHAPTER II

Of the Making of Books
There Is No End

The Rule required monks to work. Yet "work" did not always mean hard labor. A monk writing in 1075 admitted, "To tell the truth, it amounts to nothing more than shelling the new beans or rooting out weeds that choke the plants in the garden; sometimes making loaves in the bakery." A few monks worked in the stone-carver's shop or the goldsmith's atelier, but the main labor of most was in the scriptorium, making books. Like its saints' relics, the reputation of its books made a monastery famous (see Plate 2). A good library and scriptorium attracted scholars in search of knowledge and novel books to copy. Learned scholars—as Gerbert would find out—drew the attention of counts, kings, and emperors, who bestowed on monks and monasteries riches and power and, most important, protection.

All this depended on the technology of book-making.

The process, as Gerbert learned it at Aurillac, started with making parchment. According to Pliny's *Natural History*, written in the first century A.D., parchment—in Latin, *pergamenum*—was invented for the king of Pergamos (modern Bergama, Turkey) to break the Egyptian monopoly on papyrus. Parchment is made from the skins of sheep or goats, or, for special books, calves or even rabbits. Skins were everywhere; the sedge used for papyrus was common only on the banks of the Nile.

Papyrus made a fine, light sheet widely used in the Roman Empire. The outer bark is scraped off and the pith sliced into thin fibers. These are laid flat in a square, a second layer is placed at right angles, and the two are pounded until the plant's gummy sap bonds the sheet together. Papal bulls—the pronouncements of the popes, sealed with a lead drop, or *bulla*—were still written on papyrus in Gerbert's time; they changed to parchment only in 1057.

But papyrus was not suitable for the books made at Aurillac. Papal bulls are long, single-sided sheets stored rolled up. Books made of papyrus were also scrolls—not the familiar, handy, block-shaped item with stiff covers and pages that turn, which was a *codex*. Scrolls have to be read from beginning to end: You can't thumb through a scroll. And you can't make a codex out of papyrus: It cracks when it's folded.

Parchment was not as simple to make as a papyrus sheet. The process was messy and time consuming, and an average codex required many skins. To make one Bible took the prepared skins of 150 sheep. A copy of Virgil's complete works took 58 skins. Sheep-raising was thus a major activity at every monastery. The more land a monastery converted into sheep meadows, the better its library.

The oldest known recipe for turning a sheepskin into parchment was written in the Italian city of Lucca at the end of the eighth century. "Place it in limewater," it says, "and leave it there for three days and extend it on a frame and scrape it on both sides with a razor and leave it to dry, then do any kind of smoothing that you want." A twelfth-century manuscript offers more precise—but still somewhat mysterious— instructions on how "to make parchment from goatskins as it is done in Bologna." This process took twenty-four days, plus drying. Both methods made acceptable parchment, though the shorter soaking time made for harder work scraping off the hair. The monks at Aurillac may have used either one or something in between.

Limewater was the key to the process. It was made by burning crushed limestone (or marble, chalk, or shells) in a kiln to make quicklime, placing that in a vat or barrel, and adding a little water. The limewater would seethe and bubble. In about ten minutes, when the fizzing stopped, it was ready to use. The monks could spread a pulp of lime-water onto the skin, then fold the skin up and set it aside for a few days. Or they might prefer to dilute the limewater until it was milky and soak the skin in it. Either way, the lime ate into the epidermis, the outer layers of the skin, loosening the hair on one side and the fat on the other.

After rinsing off the lime, the monks pulled or plucked out whatever hair they could. They used gloves, since any remaining lime would eat into their skin, too. Next they laid the skin over a log or trestle and rubbed it with a wooden hone, a bone spatula, or a dull knife, a procedure called *scudding*.

To remove the fat, they dunked the hairless skin in fresh limewater or rubbed it with lime powder. They spread it on the trestle, hair-side down, and scudded again. Lean sheep made better parchment than fat ones; excess fat made the parchment slippery and the ink would not stick. On the other hand, parchment that was scraped too thin could become wrinkly and transparent.

The dehaired and defatted skin was soaked again, then stretched onto a wooden frame to dry. The monks wrapped each corner of the skin around a pebble (called a "pippin"), tying one end of a cord around the pippin and another to a wooden peg in the frame. This kept the cords from tearing holes in the skin. As the skin dried, the monks tightened the pegs to keep it from wrinkling. While leather-making was a chemical process, parchment-making was a physical process. What was left after all the soaking and scudding was mostly collagen, long spiraling proteins that form tough, elastic fibers. As the skin dried, these fibers tried to shrink. Stopped by the frame, instead the fibers' structure began to change.

Finally, the dried parchment was scraped again, while still on the frame, with a crescent-shaped blade called a *lunellum* ("little moon"). It was powdered again with lime or chalk to bleach it, and rubbed thoroughly on both sides with a pumice stone to raise a nap and better take the ink. The final color of the parchment depended partly on the process and partly on the animal it came from. Low-grade parchment could be dark pinkish-brown with a chalky surface, peppered with hair follicles, streaked with scrape marks, or so thin the ink bled through. Sheepskin, well cured, was yellowish, but still sometimes greasy or shiny. Goatskin was greyer. Calfskin was the whitest, though the veins could be prominent, and spots on the animal would leave spots on the parchment.

Taken down from the frame, the parchment was then cut into sheets of a standard size. The first sheet was easy: a rectangle that, folded, could become four pages of a large *quarto* book or eight pages of a small *octavo* one. Then the cutter had to become creative: A sheepskin is not square. Where the head, legs, and tail were cut off, the skin curved. Manuscript pages often have a corner missing—where a page ran into a neck hole. Other blemishes are insect bites (little holes), wounds on the animal (bigger holes), and gashes (where the knife slipped during the flaying); some of these were sewn up, but usually the scribe just wrote around them. The scraps left after the pages had been cut out were collected for rough drafts or occasional writings, like letters or wills or bills of sale, that were not bound.

While the parchment was being made, other monks prepared the inks. Black ink was made from oak galls—the black bubbles on an oak twig where the gall wasp has stung it to lay its eggs. The galls were ground, cooked in wine, and mixed with iron sulfate and gum arabic. Oak galls contain gallic acid, which causes collagen to contract; instead of sitting on the surface, the ink etched the words into the parchment. Crushed iron sulfate (often found together with pyrite) made the ink black; gum arabic, the sap of the acacia tree, made it thick. Another recipe called for vinegar and rind of pomegranate. Egg white (preserved by a sprig of cloves) and fish glue were used to further thicken inks. Other common recipes called for ear wax, pine rosin, lye, stale urine, and horse dung.

Black ink sufficed for the text of most books. But titles needed to be in red, while the lavish illuminations that graced deluxe manuscripts called for a full palette. To make red ink, the monks ground and cooked "flake-white," a white crust that formed on lead sheets hung above a pot of simmering wine. To make green, they needed copper filings, egg yolks, quicklime, tartar sediment, common salt, strong vinegar, and boys' urine. An expensive blue was made of ground lapis lazuli; a cheaper variety could be made from the woad plant, which contains the same chemical as indigo. Yellows were made from the weld plant, saffron, or unripe buckthorn berries. Purple came from the herb turnsole, brick red from madder root, and pink from brazilwood (imported from Asia), while earth colors came from filtered and roasted dirt.

The parchment made, the ink mixed, the scribe could set to work. First, using a straightedge and a knife, awl, compass, or small spiked wheel, he would prick tiny holes up the left and right margins of the page. Connecting these dots with a drypoint of metal or bone would give him lines to guide his writing. In a poor monastery—or as a beginner, like Gerbert—the scribe worked on a lapboard. Better-equipped scribes, as can be seen in manuscript illuminations, had tilted writing desks, with slots for quill pens and an inkhorn, and a ledge on which the scribe could prop open his exemplar, the book he was intending to copy. An expert scribe could write at an average rate of forty strokes (five to six words) a minute, or twenty-five lines per hour, which over an eight-hour day adds up to two hundred lines. A 180-page book, of some 5,300 lines, took one scribe (as he bragged in a colophon at the end) an entire month to copy.

And that was just for the text in black ink. A rubricator would then add chapter and section titles in red, where the first scribe indicated they should go. Illustrations and the fancy initial capitals that began major sections were usually done by a different monk who specialized in drawing. Important books were often team efforts: The manuscript known as the Harley psalter, a deluxe book of psalms, was the work of three scribes, a rubricator, and nine artists.

Finally the finished pages were sewn into quires (usually gathered together with several other books) and bound between boards of oak or beech. The inside of each board was covered with a flyleaf or pastedown, a piece of fresh parchment, or, more often, one recycled from an unwanted manuscript. Sometimes these old sheets were cleaned first, by soaking them in whey or orange juice and scraping off the inks and colors; fortunately for us, this was not always done—more than one precious leaf has been saved because it was recycled.

The outsides of the boards were covered in leather (alum-tawed pig's leather was preferred, because it was white) and fitted with metal clasps to keep the book from popping open. Sometimes the cover was ornamented with gold and gems and ivory carvings; sometimes it was left plain.

Finally, the finished book would be locked away in a wooden bookchest to protect it—not from thieves, who could break open the chest with an axe, so much as from borrowers who might "forget" to return it. For a book represented weeks of labor. In the mid-800s, Regimbert, a monk at Reichenau, bought a lawbook for eight *denaris*: the price of ninety-six two-pound loaves of bread. To a monastery, a book was a very valuable thing.

At Aurillac, Gerbert—still a penniless young student—gained the love of books that would make him one of the greatest book collectors of his age. His letters from later years are filled with requests for particular manuscripts:

> Gerbert, formerly a teacher, sends greetings to his Ayrard. We give assent to your requests, and we advise that you carry out our business as if it were your own. Have Pliny corrected; let us receive Eugraphius; and have copied those books which are at Orbais and at Saint-Basle.
>
> . . . Procure the *Historia* of Julius Caesar from Lord Adso, abbot of Montier-en-Der, to be copied again for us in order that you may have whichever [books] are ours at Reims, and may expect ones that we have since discovered [at Bobbio], namely eight volumes: Boethius *On Astrology*, also some beautiful figures of geometry, and others no less worthy of being admired.
>
> . . . You know with what zeal I am everywhere collecting copies of books. You know also how many copyists there are here and there in the cities and countryside of Italy. Act, therefore, and without confiding in anyone, have copied for me at your expense M. Man-

lius *On Astrology*, Victorius *On Rhetoric*, and Demosthenes *On the Diseases of the Eye*.

> . . . I am diligently forming a library. And, just as a short time ago in Rome and in other parts of Italy, and in Germany also, and in Lorraine, I used large sums of money to pay copyists and to acquire copies of authors, permit me to beg that this be done likewise in your locality and through your efforts so that I may be aided by the kindness and zeal of friends who are my compatriots. The writers whom we wish to have copied we shall indicate at the end of this letter.

The list of works Gerbert wished to have copied has been lost from this last letter, but some of Gerbert's books still exist. They are simple books, unadorned by colorful illuminations—copies to be read and pondered, not merely looked at. For Gerbert, books were not an entertainment, but the essential key to wisdom.

The university library in Erlangen, Germany, has a manuscript containing Priscian's *On Weights and Measures* bound together with Cicero's *On the Art of Speaking*. These were apparently copied for Gerbert by "his Ayrard," for on the last page of the Priscian, Ayrard signed his work: "Ayrard wrote this to please the venerable abbot, Gerbert the Philosopher." Notes in the margins show the note-taker to be a master of Latin style, as well as a mathematician. On the bottom of one page is a tiny signature, worn to near invisibility by age: *Gerbertus*.

The national library in Paris has Gerbert's copy of Virgil—containing the *Eclogues*, the *Georgics*, and the *Aeneid*—with notes in the margins, glosses between the lines, and a full set of teaching aids at the back. An inscription reads: "The present book reflects to the world the glory of Gerbert. It was from him alone that the bookchests of our library owe this contribution."

Gerbert apparently never requested sacred literature, only the classics, and his letters show the influence of his books. Though he quotes the Bible aptly, he is more likely to cite Virgil, Terence, or Horace. Cicero and Boethius were favorites; among the others he mentions are

Calcidius, Cassiodorus, Claudian, Eugraphius, Isidore of Seville, Livy, Lucan, Macrobius, Ovid, Pliny, Quintilian, Sallust, Seneca, Suetonius, and Tranquillus.

Raymond of Laval, the schoolmaster at Aurillac, taught Gerbert to read and write. In Gerbert's words, Raymond "shone with the double light of religion and science." We know nothing about Raymond's teaching methods, except that they were effective.

Perhaps Raymond's relationship with Gerbert was like Thangmar of Hildesheim's with his protégé Bernward, who was about ten years younger than Gerbert; both Bernward and Gerbert would later become tutors to the young Emperor Otto III. Wrote Thangmar of Bernward, "I found him of extraordinary intelligence. He would sit at the back of the room, listening with all eagerness; then, happy in his understanding, he would explain secretly to other boys what he had learned from me." Bernward's education spilled over the classtime. When Thangmar rode out on an errand for his bishop, he would take Bernward along. "Often all day long we would read as we rode, or amuse ourselves by making up verses in meter, or argue with one another some tricky point of rhetoric or dialectical debate."

By the time Gerbert left Aurillac in 967 as an adolescent (which at the time meant someone between fourteen and twenty-one), he was known for his sophisticated Latin, both written and spoken. His wit and learning would make counts, kings, and emperors overlook his humble birth and retain him as secretary, to write their letters, as schoolmaster to their sons, and ultimately as their own closest counselors. It was Raymond, Gerbert wrote, "to whom I owe everything. For [he] is an excellent teacher, especially learned, and very closely joined to me in friendship." Later, announcing to the monks of Aurillac his spectacular election as archbishop of Reims, Gerbert wrote, "To all of you together I render thanks for my installation, but more especially to Father Raymond whom, after God, I thank above all mortals for whatever knowledge I possess." He summed up his thoughts in a crisp, well-balanced Latin sentence: *Discipuli victoria, magistri est gloria,* "The pupil's victory is the master's glory."

Gerbert's education most likely began with reading the psalms. Since Charlemagne systemized the teaching in monastery schools in the late eighth century, the psalms had been the basis of every monk's education. Over the course of a week, the monks in their seven daily services would chant the entire cycle of 150 psalms in Latin. Since students were birched for their mistakes (the preferred whipping tool was a birch rod), Gerbert would have them memorized before he was old enough to face them on the page. *Beatus vir*, "Blessed is the man," he would spell out as soon as he'd learned his ABCs. As the rest of the psalm rolled off his tongue, the link between sound and symbol would become clear.

Next Gerbert learned to shape the letters. Students practiced by scratching onto wax—a wax tablet being the notebook of the time. A thin board with a rim on all four sides made a frame into which beeswax (mixed with pitch to make it black) was poured. Wax tablets were made in a handy size for hanging from one's belt. Two or more tablets could be linked with sinew along one edge and turned like the pages of a book.

Gerbert's first pen was a stylus made of wood or bone, one end sharp for scratching into the wax, the other flat like a spatula for smoothing out mistakes. From there he advanced to a quill. These were plucked from live geese in the springtime. The best were the long primaries from the tip of the left wing, which curve away from the eyes and fit comfortably in the right hand. With a penknife, he would first cut away the side barbs and slice off the end of the quill at a sharp angle. Then he would shape the nib, with curving sides and a flat tip, and test it by writing (often on the flyleaf in the back of a manuscript) *Beatus vir*.

A well-cut quill pen slid easily down the page, especially when writing on a tilted desk. But, because of the design of the nib, it always caught on the upward strokes. For this reason, the common script of the day, Carolingian minuscule, made letters with only downward strokes. Not until the fifteenth century did writers learn to split the tip of the nib lengthwise, making upward strokes possible. Angular and cramped, Carolingian minuscule can be difficult to decipher, but it was efficient: Most letters take only three quick strokes.

In Gospel books, the evangelists were usually portrayed as monks at
their writing desks taking dictation from on high. Pictured here is
Saint Luke, from the tenth-century evangelary by Bernward of
Hildesheim, Gerbert's contemporary.

Still, it was tiring. The writer had to dip his quill in the inkpot after
every fifth letter, and he had to keep his hand suspended above the
parchment while he wrote or the ink would smear. "He who cannot
write thinks it is not a labor," complained one monk in the margin of
a manuscript. "Three fingers hold the pen, but the whole body aches."
Another concurred: "It is excessive drudgery. It crooks your back, dims
your sight, twists your stomach and sides. Pray, then, you who read
this book, pray for poor Ralph." To keep his aching, sore-eyed pupils
on task, the teacher had them copy such exhortations as: "Learn how
to write, boy, so that you are not mocked. Who knows not how to
write is a living ass." Gerbert's penmanship must have been beautiful:
He wrote letters that kings and emperors were not ashamed to sign as
their own. Nor did he ever complain about the pain or drudgery of
writing.

Gerbert's next step was to learn Latin, in which he also excelled. In the
900s, the common spoken tongue was a dialect of Romance, a lost lan-

guage based on the Latin of the late Roman Empire. Aurillac's Romance was on its way to becoming French, while other forms would end up as Spanish and Italian. A good linguist—such as an abbot who entertained many travelers—could speak all three; some even spoke English or German. Latin, however, gave a monk access to any church or court in Europe, and to almost all of the books. Thanks to Alcuin of York, the English monk who organized Charlemagne's schools in the 780s, it was taught with an archaic pronunciation that accentuated the difference between it and common Romance.

Latin was not easy to learn. Medieval churchmen often garbled the rites. *In nomine patria et filia et spiritus sancti,* said a priest at baptism: "In the name of the fatherland, the daughter, and the Holy Spirit." Another prayed for "male and female mules" (*mulis et mulabus*) instead of servants (*famulis et famulabus*). Such faults were so prevalent they were given their own devil, Tutivillius. He was said to get fat off the fragments of speech that dropped from the lips of "dangling, leaping, dragging, mumbling, fore-skipping, and overleaping monks." He would produce them on the Day of Judgment to be set on the scales of sin.

Latin was likewise not easy to teach, given the tools at hand. The textbooks, such as Donatus's *The Minor Arts*, written in the fourth century, were gloriously outdated. What was obvious to Donatus, for whom Latin was his mother tongue, was not so obvious to Sedulius the Scot, who taught it in the 800s. Donatus's simple comment that there were eight parts of speech took Sedulius fifty-five lines to explain. Finishing a copy of Priscian's *Introduction to Grammar*, one scribe despaired. Latin grammar, he wrote, was "like a sea without a shore. Once you fall in, you rise no more."

To improve their teaching, some schoolmasters created their own textbooks. In the tenth century, Egbert of Liege turned hundreds of fables and proverbs into memorable little ditties in Latin hexameter verse, including the earliest version of "Little Red Riding Hood." Aelfric of Eynsham wrote his *Colloquy* in two languages, Old English and Latin. Framed as a dialogue between teacher and pupils, the *Colloquy* talks about the ordinary things in a young monk's life, giving him the

Latin words for the work of the cook and the baker, the plowman and the smith, the shepherd and the milkmaid. There are bilingual lists of what the hunter hunts (with hawk or hound), what the fisherman catches, and what the gardener prunes and weeds. The merchant is quizzed about crossing the sea; the salter explains how, without him, meat, butter, and cheese would not keep.

The poets provided another door into Latin—particularly the jar of snakes that was Virgil's *Aeneid*. Though eleventh-century writers would warn that too great a passion for pagan authors led to heresy, in Gerbert's day the classics were considered indispensable for learning how to write and speak well. As Gerbert later wrote to a friend, "Nothing in human affairs is more worthy of veneration than the wisdom of famous men which is contained in the multitudinous volumes of their books. Continue therefore as you have started, and quench your thirst in the waters of Cicero."

To write or speak eloquently, memorization was key. Since books were rare—even the great library at Bobbio, with its 690 books, did not have classroom copies of Virgil's *Aeneid*—the schoolmaster would read a poem aloud and explicate it line by line. The students were expected to learn it by heart (or at least to learn certain parts) and to be able to slip a reference elegantly into an argument, whether spoken or written. Other teachers, such as the ninth-century master Lupus of Ferrieres, compiled books of pithy sayings on such topics as miracles, visions, war, abstinence, constancy, poverty, love, old age, parents, and patience. These, too, the students were to memorize and cite when appropriate.

Gerbert became quite skilled at this trick, sprinkling his letters with classical allusions. Twice, for example, he quotes Virgil's *Aeneid*—"The features of my friend remain fixed in my heart"—to describe his sorrow at their parting. He concludes a letter asking for books with the words of Cicero: "Equally in leisure and in work we both teach what we know, and learn what we do not know." Rarely does he identify the author he is quoting. To his favorite student he writes about a rival teacher: "Though really still a learner along with me, he pretends that only he has knowledge of it, as Horace says." More often, Gerbert assumes his reader, as a learned man or woman, will recognize and appreciate the allusion.

Like his contemporaries, Gerbert did not use a notebook to store these favorite sayings or bits of poems. Instead, he created a "house of memory": He built in his mind an imaginary palace of many rooms. When he needed to speak on the subject of "patience," he would mentally walk down the corridors and up the stairs until he reached the little turret room labeled *patience*, and there he would find the *mot juste*.

But stock phrases, figures of speech, rhymes, allusions, and other rhetorical flourishes were of secondary importance to the art of discourse itself. As good grammar begins with subject and verb, eloquence begins with exhortation, narration, argumentation, refutation, and conclusion. The goal of rhetoric was to speak and write with elegance and restraint, and while many tenth-century writers forgot those rules and prattled on in wordy, self-aggrandizing pomp, Gerbert's writing was always to the point. Though he was somewhat prone to exaggeration, his style was consistently simple, clear, well-organized, and direct, even when he was being sarcastic, sycophantic, or purposely misleading. He displayed, said a tenth-century colleague, "Ciceronian eloquence."

From rhetoric, Gerbert advanced to dialectic. Both are kinds of logic, says a poem by Fulbert, the schoolmaster of Chartres in 1004. (Fulbert may have been one of Gerbert's own students at Reims.) Rhetoric, Fulbert explains, uses "flowing speeches" to persuade. Dialectic takes the question-and-answer form of a dialogue; its goal is to "compel one's opponent to concede." Its primary tool is the syllogism. The flyleaves of manuscripts, again, were used for practice. On one, for instance, a student tested out his argument that God is not "anywhere," because only bodies can be contained in places, and since God does not have a body, he cannot be "anywhere."

As rhetoric is associated with Cicero, dialectic is linked to Aristotle—and thus to Boethius. Boethius's works so influenced Gerbert that, in

997, he convinced the emperor to erect a monument to "the father and light of the world." The statue no longer exists (if it was ever made), but Gerbert's inscription does. Of Boethius, he wrote: "You shine light on knowledge and you need not yield to the talents of the Greeks. Your divine mind keeps in order the power of the world."

Boethius was born soon after the fall of the Roman Empire in 476 A.D. When the empire was divided into East and West, the Roman elite were no longer educated in Greece. The classics of philosophy, medicine, mathematics, and science—all in Greek—were no longer read. Boethius, realizing this as a student, dedicated himself "so far as life and leisure for work are vouchsafed me" to translating into Latin all of Aristotle's works. While serving as adviser to Theodoric, king of the Ostrogoths and ruler of the western Roman Empire, Boethius translated the six books of logic and a popular textbook, Porphyry's *Introduction to the Categories of Aristotle*. He wrote five commentaries on Aristotle and one on Cicero, as well as textbooks on arithmetic, music, and astronomy. Then he was arrested on a trumped-up charge of treason. He was imprisoned and, in 525, bludgeoned to death.

While in prison in the Italian city of Pavia, Boethius wrote his most famous book, *The Consolation of Philosophy*. Philosophy in this work is personified as a woman of "awe-inspiring appearance, her eyes burning and keen beyond the usual power of men." Using dialectic, she leads Boethius to the logical conclusion that virtue is its own reward. Fortune notes that, if "the things whose loss you are bemoaning were really yours, you could never have lost them." Philosophy then proves that "goodness cannot be removed from those who are good." Furthermore, "perfect good is true happiness," and "true happiness is to be found in the supreme God"; "each happy individual is therefore divine."

Faced with his own political setbacks, Gerbert often thought of Boethius and his *Consolation*. As he wrote to his teacher Raymond, "For these cares philosophy alone has been found the only remedy." He was impressed not only by Boethius's conclusions, but by his logic.

After studying the *Consolation* and other written dialogues by Boethius and Aristotle, students of dialectic advanced to real debates—

sometimes with a "sophist," a professional debater, brought in for the purpose. A description of Gerbert's own classroom, when he was schoolmaster at Reims, explains: "They engaged in practice disputation so that their speech might seem artless, as becomes those who are masters of their art."

Gerbert knew logic was not enough to win an argument. According to Charlemagne's schoolmaster, Alcuin, delivery was crucial. The people of Gerbert's time, as in Alcuin's, believed the body mirrored the soul. How one walked and gestured, the angle of the head, the set of the shoulders, one's grave, modest, and, above all, *appropriate* expressions—these were the outward signs of virtue. And whereas Cicero argued that virtue made an orator good, to Alcuin it made him effective. The cumulative power of charm, grace, and wit, what he termed "nobility of soul," could even outweigh nobility of blood. And so it did in Gerbert's case.

Grammar, rhetoric, and dialectic were only the first three of the seven liberal arts. Known as the *trivium*, the threesome, they were considered elementary—though many monks never advanced beyond grammar, and even Raymond of Laval, Gerbert's teacher at Aurillac, was not skilled in dialectic.

The *quadrivium* was a foursome of number-related disciplines: arithmetic, geometry, astronomy, and one that comes as a surprise to us today, music. Far from the practical studies of accounting, architecture, timekeeping, or chant, which were learned by doing, the quadrivium was an exploration of the mind of God, who, according to the Book of Wisdom (which is still in the Roman Catholic and Eastern Orthodox versions of the Old Testament, but not in the Protestant Bible), had ordered the world according to number, measure, and weight. Adding trivium and quadrivium together, the seven liberal arts prepared a monk for a lifelong study of philosophy (some said) or theology (others insisted), or, as Gerbert himself phrased it, "the understanding of all things human and divine."

There was no one at Aurillac, or anywhere in France, who taught the quadrivium in 967, when Gerbert was ready to advance. To continue his education, he would have to go elsewhere. We can't say it was all his idea, as an independent-minded adolescent, to leave his monastery and go off to learn the mathematical arts. The legend says he "made his escape one night." But the historical accounts (and his letters home) show he had Abbot Gerald's permission to go. Regardless, at seventeen or perhaps a little older, Gerbert of Aurillac was let out into the world. He would never live as a cloistered monk again.

CHAPTER III

The Ornament of the World

Count Borrell of Barcelona came to Aurillac in 967. He had crossed the Pyrenees to wed the Lady of Rodez and continued north some 50 miles to kneel before the miracle-working bones of Count Gerald the Good. The abbot welcomed him to the monastery's guest hall, and the two sat down to talk. Their conversation was reported by the monk Richer of Saint-Remy, who knew Gerbert well in later years and must have heard him talk about it:

> The abbot inquired whether Spain contained any scholars very learned in the arts. When the count very promptly asserted that there were, he was at once persuaded by the abbot to take one of the monks with him for the purpose of having him further instructed. Therefore, since the count had no objections, he freely granted the favor, and with the unanimous consent of the brothers took Gerbert as their choice, and turned him over to Bishop Ato to be taught, with whom he studied mathematics extensively and successfully.

Gerbert's three years in Spain, from 967 to 970, were the defining episode of his life. He not only mastered the quadrivium, he made life-long friends who shared his interest in math and science. He also learned how a kingdom could be run—and what part a churchman and scholar could best play in it. In later years, he would look back longingly on this time. He discussed Count Borrell's fortunes with the abbot of Aurillac in 986 and, in spite of the sacking of Barcelona the year before, told another correspondent that he expected Spain, "which has been neglected for a long time, will seek me again." As pope, he took special care of the Spanish churches; when he died, in 1003, he was mourned by Count Borrell's second son, who marked the passing of the "glorious and very wise Pope Gerbert."

Count Borrell's Spain straddled the Pyrenees and progressed south to al-Andalus, the Islamic caliphate that controlled most of the Iberian Peninsula. To Borrell's west were the Christian kingdoms of Leon, Castille, and Navarre; but Borrell's country (roughly what is now Catalonia) was officially part of France. Culturally, however, France and Catalonia were quite different.

Law, rather than counts or castellans, ruled in Catalonia. Unlike in Aurillac, where "laws" applied only when a band of knights enforced them, in Catalonia disputes were settled in public courts, with formal procedures overseen by professional judges. Justice was an analytical process—these judges tried to "find the law" in a case. They were not mere mediators, in search of a truce between aggrieved parties. They called witnesses, heard testimony, and even accepted written evidence—for Catalonia was an unusually literate society. Deeds, wills, titles, sales, and all sorts of routine transactions were written down on scraps of parchment and squirreled away in coffers and archives to be brandished at need before a judge's bench. Noblemen and churchmen, as well as peasants, could be summoned by a judge, and all agreed it was the count's duty to see that the law remained unsullied by bribery or threats.

This law derived from that of the Visigoths, whose king in 654 issued the most sophisticated code of law the medieval world had yet seen. The Visigoths, a Germanic tribe, are remembered for having sacked Rome in 410 and conquered Spain in the 470s. But they were not all warriors. In addition to preserving Roman law, they prized Latin learning: Isidore of Seville wrote his encyclopedia for the Visigothic king in the early 600s; copies were still being made in Catalonia (and throughout Europe) in Gerbert's day.

The Visigothic kingdom of Spain lasted until 711. That year, while the king was far in the north trying to subdue the Basques, a Muslim force invaded from the south. Fleet after fleet crossed the Mediterranean from the Maghrib. Meeting little resistance, the Muslim armies pressed northward. In 732, they were finally stopped by the French 150 miles from Paris. They were forced back to Provence, and then south of the Pyrenees, where they remained for more than 700 years: The last Muslim ruler in Spain was not defeated until 1492.

The invasion of 711 was impressive. Some 150,000 to 200,000 Muslim warriors, joined by their wives, children, and slaves, eventually settled in the Iberian Peninsula. But while they shared a religion, these newcomers were not united. The majority Berbers chafed under the rule of their Arab overlords, and the Arab leaders quarreled among themselves. The kingdom would probably have collapsed if Abd al-Rahman had not arrived in 755. Son of an Arab father and a Berber mother, he was the only surviving member of the Ummayad clan. The Ummayads had ruled the House of Islam until the rival Abbasids invited them all to a dinner party in Damascus and murdered them. Only Abd al-Rahman, then a teenager, escaped. He fled to Spain, where, inspired by his hereditary right to kingship, he gathered an army of his own and conquered Cordoba. Meanwhile, the Abbasids built Baghdad, the City of Peace, and moved the capital of the Islamic empire there.

Twenty-two years later, Abd al-Rahman was still fighting his way back to the Pyrenees. His enemies, until then, had been other Arabs, proud Muslim lords unwilling to submit to an overlord, no matter how royal. Now, with their backs to the mountains, these lords looked north for reinforcements, and they called for aid from Charlemagne.

The king of the Franks and his new Arab allies battled Abd al-Rahman from Barcelona to Pamplona. In 778, defeated, Charlemagne began to retreat—several hundred years later, the story became the basis for *The Song of Roland*, the first French epic. At Roncesvalles in the Pyrenees, the poem says, the weary and demoralized Franks were attacked from the rear. The Christian hero Roland, trapped and outnumbered by Muslim enemies, bravely tried to hold them off. Finally, knowing he had failed, he blew his horn to call for help. Too late, Charlemagne turned: His rearguard was defeated, slaughtered to the last man. But *The Song of Roland* was not written until the eleventh century. The poem transformed what really happened into a completely unrelated story of Christian versus Muslim, infused with the religious hatred that sparked the First Crusade in 1096. Abd al-Rahman's Muslims did not attack and kill Roland and his faithful friend Oliver. Nor did Charlemagne's own Muslim allies turn on them. The killers were the Christian Basques. The conflict in Spain in 778, as Gerbert would have known, was over territory, not faith.

Al-Andalus was not challenged by Franks for another hundred years. Then, in 878, the wonderfully named Guifre the Hairy received the title of Count of Barcelona from the French king. From his base in Toulouse, Guifre swept the Arabs southward. Taking control of the lands that would become modern-day Catalonia, he encouraged the remnants of the Visigoths to come down from the mountains, where they had hidden for centuries, and resettle the plains. He built castles and established monasteries and brought back the Law of the Goths.

Count Guifre decided unilaterally that his title was hereditary. His sons and nephews became counts, parceling out the property he had conquered. His grandson, the Borrell who had brought Gerbert to Spain, gathered together the counties of Barcelona, Girona, Vic, and Urgell and called himself duke of the Goths (a title the count of Toulouse also affected). The neighboring counts were Borrell's cousins, while the youngest members of the clan were appointed abbots and bishops— thus blurring the lines between church and state. Elsewhere, such arrangements were seen as corrupt; here, they were considered efficient: Borrell's kinsmen were talented and well-educated men.

In Catalonia, Gerbert learned lessons he would later apply in his dealings with kings. Borrell's relationship with Bishop Ato, Gerbert's supervisor, for example, showed him how church and state, both bound by law, could work in harmony for the greater good. Though not a kinsman, Ato held for Borrell the frontier town of Vic, just west of Barcelona. An ambitious man, he was not content to be a mere bishop. He saw Vic as the logical site of an archbishopric—and himself as the natural overseer of all churches and churchmen in Catalonia. There was, at the time, no archbishop south of the Pyrenees. When the previous archbishopric, Tarragona, had been recaptured by the Muslims twenty-some years before, Rome had, strangely enough, given the power of oversight to the archbishop of Narbonne. That city was so far on the French side of the Pyrenees as to be considered "foreign" to Borrell's Catalonians. While Gerbert was in Spain, Borrell and Ato were devising a scheme to correct the pope's error and bring the power of the Catalan church back under local control.

Yet even as a mere bishop, Ato was a powerful lord. He had the right to oversee the judicial courts in Vic, keeping all the fines for himself. He pocketed the tolls on roads and bridges and the taxes from town markets. He could mint his own coins. Along with overseeing the borderland's churches and monasteries, he controlled several castles—at one point the total was twenty-five—positioned to protect the outlying settlements. He was essentially a knight. No one except the monk Richer of Saint-Remy described him as a mathematician. But whether he was interested in numbers or not, he was very well-connected. With such a mentor as Ato of Vic, Gerbert had found the perfect place—on the border of al-Andalus—to study the mathematical arts.

Arabic was then the language of science. It was from al-Andalus, beginning in Gerbert's lifetime, that the essence of modern mathematics, astronomy, physics, medicine, philosophy—even computer science—would seep northward into Christian Europe over the next three hundred years.

This science, in general, came from Baghdad. Through the tales of Scheherazade, *The Arabian Nights*, the Abbasid Empire has entered our popular culture. We all know of Aladdin and his magic lamp, the flying carpet, "open sesame." We may even remember the caliph Harun al-Rashid, who appears in many of the tales. But Scheherazade does not mention his House of Wisdom, with its staff of translators, scribes, and bookbinders. Harun al-Rashid's son and successor, al-Mamun, expanded the program in the early 800s, turning it into a sort of research institute in mathematics, astronomy, and medicine.

Scholars from the House of Wisdom went looking for old books in Persian, Sanskrit, Syriac, and particularly Greek. When Hunayn ibn Ishaq, the chief of translation from 847 to 861, wanted to read a medical text written by Galen in the second century A.D., he set out to find it: "I myself searched with great zeal in quest of this book over Mesopotamia, all of Syria, in Palestine and Egypt until I came to Alexandria," he wrote. He found a partial copy in Damascus, "but what I found was neither successive chapters, nor complete." A friend, however, found additional chapters elsewhere, and the book was pieced together and translated into Arabic.

In this way Baghdad's House of Wisdom saved works by Euclid, Archimedes, Aristotle, Ptolemy, and many other classical thinkers. The Greek versions have disappeared. Only because the Arabic versions survived to be copied and, eventually, translated into Latin was knowledge not lost.

Thabit ibn Qurra, who lived until 901, was one of the translators of Galen's book. He is known to have translated nearly two hundred books of medicine, mathematics, and astronomy. Thabit also did original research: He showed how algebraic and geometrical proofs were related to each other, for example. This was the subject, applied to triangles, that Gerbert would explore with his friend Adalbold in 999.

Of the mathematicians who frequented Baghdad's House of Wisdom, however, Muhammad ibn Musa al-Khwarizmi, who died in 850, was chief. He wrote the first book on what we call Arabic numerals: He named it *On Indian Calculation*, well aware that the symbols 1 to 9, and the place-value system that makes arithmetic easy, originally came from India. Modern algebra (from the Arabic *al-jabr*) comes from his book *Kitab al-muktasar fi hisab al-jabr wa'l-muqabalah* (The Compendious Book on Calculation by Completing and Balancing). The word *algorithm*, without which no computer scientist could function, is derived from al-Khwarizmi's name.

Near the House of Wisdom was an observatory, for the stars were central to the caliphs' interests. A third influential book by al-Khwarizmi was his *Zij al-Sindhind*, or Star Tables: several hundred pages of text and tables in which he uses trigonometry, spherical astronomy, and other advanced math to calculate, for specific locations and dates, the changing positions in the heavens of the planets, sun, moon, and stars.

The caliph al-Mamun sponsored al-Khwarizmi's science, in part, because he liked to have his horoscope read. He also wanted an accurate map showing the full extent of his empire, which required al-Khwarizmi to calculate the circumference of the earth. In 827, the caliph sent two parties of surveyors to the plain of Sinjar, 70 miles west of Mosul. One group walked north, the other south. Two hundred years later, the Persian astronomer al-Biruni described their experiment, which used a method devised by Ptolemy. "Each party observed the meridian altitude of the sun until they found that the change in it amounted to one degree," he wrote. As they walked, they "planted arrows at different stages on their paths" to measure the distance. On the way back to their starting point, they took a second set of measurements to doublecheck the first. One degree, they determined, measured 64.5 miles. Multiplying that number by 360 degrees, they found the circumference of the earth to be 23,220 miles—not far from the actual circumference, which is 24,900.

The caliphs' surveyors measured the sun's altitude with an astrolabe, an instrument said to have a thousand uses, such as locating a star, telling time, finding the direction of Mecca or the height of a tower, surveying land, and casting a horoscope. The astrolabe was the most popular scientific instrument of the Middle Ages, and al-Khwarizmi was one of the first in a long line of scholars—including Gerbert or one of his students, and culminating with Chaucer more than five hundred years later, in 1391—to write a treatise explaining how to use one (see Plate 3).

Using the measured result of 64.5 miles for one degree of longitude, al-Khwarizmi made the caliph a map of the known world. He used

Ptolemy's methods for mapping a sphere onto a flat surface (the same math needed to construct an astrolabe). He then calculated the latitude and longitude of important places on the earth, using his observations of the stars. He checked his results against reports from travelers on the time it took to get from one place to another. Finally, he shaded in the extent of the Abbasid Empire.

Al-Khwarizmi's science quickly reached Islamic Spain—perhaps even before his death in 850. By the tenth century, with the Mediterranean islands of Crete and Sicily under Muslim control, trade flourished between East and West. Spain sold figs to Baghdad and imported copper pots from India. (It also sent those pots back to India to be repaired.) Spanish merchants sold antimony in Aden and bought pepper and flax. From China they imported porcelain bowls.

In 948 the Arab traveler Muhammad ibn Hauqal was working on his *Description of the World.* Al-Andalus was a magnificent land, he wrote, "of forests and fruit trees and rivers of sweet water." Most of it was cultivated and well-settled. "Abundance and ease" were the dominant aspects of life there, and a great deal of gold was in circulation. Spanish fabrics—linen, wool, silk, brocade, and "the most beautiful velvet you can imagine"—were sought after in Egypt, Mecca, and Yemen. Spain raised the best mules in the world and sold the best slaves. In particular, it was the source of "all the eunuch Slavs found on the face of the earth," these having been brought south by German and French merchants and castrated in Cordoba by Jewish doctors who specialized in the practice.

The Jewish vizier Hasdai ibn Shaprut described his country in a letter at about the same time. Al-Andalus "is rich, abounding in rivers, springs, and aqueducts; a land of grain, oil, and wine," he wrote. Among its resources were "the leaves of the tree upon which the silkworm feeds, of which we have great abundance," as well as cochineal and crocus, for dying the cloth. Silver, gold, copper, iron, tin, lead, sulphur, porphyry, marble, and crystal, he added, were mined in the mountains.

Other resources of Spain are described in the *Calendar of Cordoba*, written in about 960. It tells when to plant and harvest such crops as sugar cane, rice, eggplant, watermelon, and banana—all of which were brought to Spain by Arab settlers. They also introduced cotton, oranges and grapefruits, lemons and limes, apricots, olives, spinach, artichokes, and hard wheat, as well as the techniques of crop rotation, fertilization, and irrigation using canals and water wheels.

Another technology Arab settlers brought west was paper-making. The surname al-Warraq, "the papermaker," was first seen in Spain in the tenth century, and a water-powered papermill was built near Valencia before the century's end. Paper, made from linen rags, was both cause and effect of the Muslim love of books. Any young man who could afford it went east for his education and brought back the latest scientific and philosophical tomes. They made quick copies on cheap paper and, once home, transferred the text to more durable parchment. It would be several hundred years before Christian scholars learned to do the same.

Some of the travelers to the east were professional bookbuyers. Abd al-Rahman III, who reigned from 912 to 961, was known as a learned man—it added to his prestige. But his son, al-Hakam II, who would have to wait until he was forty-five to succeed his long-lived father, was the true scholar. With the wealth of Spain at his disposal, he employed bookbuyers in every Muslim land, as well as a team of copyists. His Royal Library in Cordoba, just west of the Great Mosque, was said to contain 400,000 books in 976. By contrast, the greatest Christian library of the time, at the monastery of Bobbio in Italy, held only 690 books.

Four hundred thousand may be an exaggeration. The catalog of the library, now lost, was said to fill forty-four books, each with a hundred pages. For the full set to contain 400,000 titles, each page would need to hold ninety titles—difficult, if not impossible. One-tenth of that number, nine titles per page, would easily fit. Even at only 40,000 books, the library of Cordoba was far and away the largest library in Europe.

Compared to the other superlatives used to describe Cordoba in the tenth century, a library of 40,000 books is not absurd. The city was nearly half as big as Baghdad, the largest city of its day. It held hundreds—maybe thousands—of mosques. Running water from aqueducts supplied nine hundred public baths. The goldfish in the palace ponds ate 12,000 loaves of bread a day. The paved streets were lit all night. The postal service used carrier pigeons. The munitions factory made 20,000 arrows a month. The market held tens of thousands of shops, including bookshops, and seventy scribes worked exclusively on producing Korans.

Cordoba impressed everyone who heard of it. In 955, the nun Hrosvit of Gandersheim met an ambassador from Cordoba at the German court of Otto the Great. She recorded in a poem what she had learned of his city. "The brilliant ornament of the world shone in the west," she wrote. "Cordoba was its name and it was wealthy and famous and known for its pleasures and resplendent in all things, and especially for its seven streams of wisdom"—these streams being the seven liberal arts: grammar, rhetoric, and dialectic; and the ones Gerbert was pursuing, arithmetic, geometry, astronomy, and music.

Significantly, only about half of Cordoba's residents were Muslim. The Koran teaches that, since Moses and Jesus had both been given books by God, Jews and Christians, like Muslims, were "People of the Book," and thus to be tolerated. In al-Andalus, this general creed of tolerance was codified in the form of a *dhimma*, a pact or covenant between the rulers and their subjects. Christians and Jews were not forced to convert to Islam, but could practice their religions—as long as they did so quietly and didn't proselytize. Other than having to pay a head tax (which Muslims did not) they were not excluded from the city's social or economic life. They could, and did, fight in the army. Depending on the ruler's interpretation of the law, and their own talents, they could advance to the highest political posts. The ambassador whom Hrosvit met in Germany, the Christian Bishop Racemundo, was one of the caliph's closest confidants (he may also have written the *Calendar of Cordoba*).

Another was the vizier, Hasdai ibn Shaprut. Prince of the Jews in al-Andalus, Hasdai is one of the most famous figures of tenth-century Spain, equally well known as a politician and an intellectual.

Gerbert arrived in Spain during the time that later poets would name its Golden Age. Arabic was the lingua franca, not just the language of religion. Christians wrote erotic poetry in Arabic; they also sang Mass in that language. They studied the latest translations sent out from Baghdad, sitting side by side with their Muslim and Jewish peers, without any suggestion they were betraying their faith. This vision of a kingdom based on religious tolerance and scholarly inquiry was the second lesson Gerbert would learn living on the border of al-Andalus.

An anecdote recorded by the Cordoban doctor Ibn Juljul in 987 gives just such a picture of the city's intellectual life. In 949 the caliph had sent Bishop Racemundo to the emperor of Constantinople. The gifts Racemundo brought home included a green onyx fountain and two books: "the book of Orosius the narrator, which is a wonderful Roman book of history, containing records of past ages and narratives concerning the early monarchs," and Dioscorides' *On Medicine*, "illustrated with wonderful pictures of the herbs in Byzantine style." A Christian scholar was given the task of translating the Latin Orosius into Arabic. But the Dioscorides was in Greek and none of the Christians of al-Andalus, Ibn Juljul noted, read Greek. The library of Cordoba had another copy of Dioscorides from Baghdad. But the translators at the House of Wisdom could not identify all of the medicinal herbs and left many plant names in the original Greek.

"There was in Cordoba a group of doctors who were keen to find out by research and inquiry the Arabic names of the simple remedies of Dioscorides that were still unknown," Ibn Juljul wrote (he would join that group in the 960s). He continued, "They were encouraged in that research by the Jew Hasdai ibn Shaprut."

The caliph sent word of their difficulty back to the emperor of Constantinople, and in 951 there arrived in Cordoba a monk named Nicholas who was fluent in both Greek and Latin. Hasdai "favored and honored" Nicholas, says Ibn Juljul, above all the rest of the group—Christians, Jews, and Muslims—who sat down together to translate *On Medicine*. Hasdai himself wrote the final Arabic version.

Though there's little record of it, we can imagine a similar collaboration of Christians, Jews, and Muslims, Arabic-speakers and Latin- or even Greek-speakers, sitting down together to translate and learn from the many books of mathematics and astronomy that came from Baghdad before the year 1000.

Writers in the eleventh and twelfth centuries credit Gerbert himself for bringing the new mathematics and astronomy north from Cordoba. *Gerbertus Latio numeros abacique figuras* runs a verse on two mathematical manuscripts: "Gerbert gave the Latin world the numbers and the figures of the abacus," meaning—as the illustrations show—the Arabic numerals 1 to 9, as explained by al-Khwarizmi. Seven manuscripts (out of eighty) give him credit for the first Latin book on the astrolabe, again based on al-Khwarizmi's work.

Even William of Malmesbury, whose twelfth-century history of Gerbert's stay in al-Andalus otherwise reads like *The Arabian Nights*, mentions these two scientific instruments. Gerbert, he says, "surpassed Ptolemy in knowledge of the astrolabe" and "was the first to seize the abacus from the Saracens." (Gerbert also learned "to interpret the song and flight of birds" and "to summon ghostly forms from the nether regions," William adds, before launching into a tale of Gerbert finding buried treasure in Rome by interpreting a statue.)

Richer of Saint-Remy, writing in the 990s, does not mention the astrolabe, though he goes into detail about Gerbert's abacus. He does not mention Saracens, only Spain and Bishop Ato. But Ademar of Chabannes, in about 1030, clearly states that Gerbert, "thirsty for knowledge," went to Cordoba.

It's possible that he did. In the cathedral treasury of Girona is an Arabic *arqueta*, an elaborate casket of gilded silver embossed with medallions of lilies (see Plate 4). It is big enough to fit two substantial books; it could also have been used as a reliquary. These boxes of ivory, wood, or precious metals were common diplomatic gifts. This one was given to Ger-

bert's patron Count Borrell by Caliph al-Hakam II sometime between
961 and 976, dates that overlap with Gerbert's stay from 967 to 970.

In fact, travel between Cordoba and Barcelona had been frequent
since 940, when a détente between the two kingdoms was brokered
(while twelve Cordoban warships blocked the harbor of Barcelona) by
Hasdai ibn Shaprut. The great Jewish intellectual spent at least four
months in Catalonia in 940; then he returned to Cordoba, shepherding
Barcelona's ambassador, a monk named Gotmar. An Arabic source says
Gotmar brought Prince al-Hakam a history he had written of the
French kings—al-Hakam's scholarly leanings were well known.

How long Gotmar remained with Hasdai in Cordoba is unknown,
as are the gifts he returned with. But if any translated books were among
them, they could have reached Gerbert. News of Cordoba's intelligentsia certainly would have, for Gotmar was named bishop of Girona
in 944. Ato was his archdeacon there until 957, when he became bishop
of Vic and, in 967, Gerbert's mentor.

Heading south from Aurillac with Count Borrell in 967, young
Gerbert may have seen Conques and the golden majesty of Saint
Foy for the first time. Doubtless their route led from monastery to
monastery. Traveling 20 to 30 miles a day (not difficult on a soft-
ambling Spanish mule), it would have taken about two weeks to reach
Vic. They rode through Rodez in its wide valley, picking up Borrell's
bride if she was not already in the party, and past Albi, its red sand-
stone towers rising bright above the river Tarn. Beyond the Black
Mountain with its deep forests, the landscape changed to a dry, windy
scrubland; a bank of hills approached, snowy mountains rearing up behind. On an old Roman bridge, well built of pink and tan stone, they
crossed the river Aude.

Entering the foothills of the Pyrenees, they squeezed through river
gorges and clung to rocky cliffs. Wine grapes grew on every possible
acre, some still watered by Roman aqueducts. Across a high pass, where
cattle grazed beside a dolman more ancient even than the aqueducts,

they saw ahead the great white face of Mount Canigou. At its foot was the monastery of Cuxa, nestled in a bowl of wooded hills. Cuxa's tile-roofed cathedral was one of the tallest churches of the tenth century. When Gerbert visited, it was still under construction; it would not be consecrated until 974. Its 130-foot belltowers were not built until the eleventh century, but the lofty grandeur of its nave would have impressed the young monk. He may even have noticed the strange keyhole shape of its archways, derived from Arabic architecture.

High in the hills above Cuxa were hot springs, popular since Roman times. Even higher was a white stone chapel dedicated to Saint Martin, with a square nave and rounded apse just like the original church at Aurillac. It clung to the side of a cliff with an ethereal view: It was called the "balcony of Canigou" until another church and monastery were built, still higher, in the eleventh century, and took over the title. Trails radiated out from each church, some marked with scallop shells as the road to Compostela: The Pyrenees were routinely crossed here. For some years, the abbot of the monastery at Cuxa was simultaneously abbot of Ripoll, on the south side of the mountains.

Coming down through thick chestnut forests, past another round-ended white church and fields full of horses, Gerbert could glimpse the Mediterranean sparkling in the distance. The way led them through a deep river gorge to the monastery of Ripoll, at the confluence of two rivers. South again, the landscape opened up, a wide fertile plain spreading 20 miles to Vic, where Gerbert would meet his new mentor.

The Romans had settled Vic and named it Ausona. But when Guifre the Hairy reclaimed the area from the Arabs in 880, he had a castle built on the site of the Roman temple and a church on the outskirts, or *vicus*, of the town. When the bishop became more powerful than the castellan, the name of the town was changed to Vic. It was a prosperous bishopric, surrounded by good agricultural land. Compared to the steep hills and deep gorges of the Pyrenees, Vic was flatter, its soil lighter, its rivers more easily tamed for gristmills and irrigation. To the southeast, the hills of Montseny blocked the Mediterranean breezes, making Vic colder and foggier than sunny Barcelona, 30 miles away. East was an-

An arch in the Arabic keyhole style at the restored cathedral of Cuxa.
The cathedral was built by Gerbert's friend Abbot Garin and
consecrated in 974.

other jagged range of hills clothed in beech woods, the rugged land beyond pockmarked with volcanic craters. Churches and monasteries
crowned many a sheer bluff, providing a fortress refuge—and a look-out
point, for from Vic the flat plain continued to the border and beyond.
No mountains blocked the armies of al-Andalus—or the merchants and
craftsmen bringing silk, gold, and science, along with the technique of
building archways in the shape of keyholes, to the north.

Vic was Bishop Ato's main residence, but Gerbert probably studied at
Cuxa and Ripoll as well during the three years he lived in Spain. There
were no scientific manuscripts at the cathedral of Vic, according to an
inventory made in 971. Its tiny library of fifty-nine books held nothing
that couldn't be found in Aurillac. Gerbert left us none of the letters he
might have written while he was in Spain—he did not begin saving

copies of his letters until 982—and no Catalan document mentions the young monk from Aurillac. But other evidence links both Cuxa and Ripoll to the study of mathematics—particularly, to Arabic mathematics.

At Cuxa, Abbot Garin's interest in Arabic science can be seen in the design of the church he built—particularly in the keyhole arches. Originally from the monastery of Cluny, Garin was abbot of four monasteries in southern France, in addition to Cuxa, when Gerbert met him; he ran them all in the reformist style of Odo of Cluny, with an emphasis on strictness and learning. He was a wily politician as well, an idealist who hoped to see the restoration of Charlemagne's empire in his lifetime. This was a dream he shared with Gerbert—and he offered Gerbert valuable insights on how to achieve it. Garin traveled to Venice at the request of the pope, to keep that vital city from allying with Constantinople instead of Rome, and brought home with him the penitent doge. Garin had convinced the Venetian ruler that betraying Rome was betraying God; the doge lived his last years at Cuxa as a hermit. A few years later, Garin took a pilgrimage to Jerusalem, taking up a collection along the way for the Church of the Holy Sepulchre. At the consecration of his Arab-influenced cathedral at Cuxa in 974, Miro Bonfill, the bishop of Girona and count of Besalu, called Garin "a dazzling star" who "shook up the world."

But Gerbert remembered Garin best as a mathematician. Many years later, in 984, he would write to Abbot Gerald of Aurillac asking for a copy of "a little book on the multiplication and division of numbers by Joseph the Wise of Spain" that Garin had left there. This "Joseph the Wise" may have been Hasdai, the caliph's vizier: Hasdai's full name was Abu *Yusuf* (Joseph) Hasdai ben Ishaq ibn Shaprut. A casual mention of him in a book by a tenth-century Jewish intellectual in Tunisia underscores Hasdai's interest in the mathematical sciences. Describing the lunar phases, the Tunisian points out: "We have explained this phenomenon and represented it in figures in our astronomical work sent to Abu Yusuf Hasdai ben Ishaq." Unfortunately no manuscripts of that astronomical work (or the "little book on the multiplication and division of numbers") have been found.

Gerbert wrote Miro Bonfill asking for this same book on numbers, and he may also have asked Garin directly for a copy. In 985, Gerbert did Garin a favor by composing his appeal for alms for the Holy Sepulchre. The same year he told his teacher Raymond at Aurillac that he was looking for a new patron and, "influenced by the encouragement of our friend Abbot Garin," was "considering approaching the princes of Spain."

The school at Ripoll was also famous in the Middle Ages as a center of learning. But the abbot Gerbert met there was best known for his military prowess: He would die leading an attack on Cordoba in 1010. Ripoll's reputation seems mostly due to the famous scholar and statesman Oliba.

Oliba was a count before he joined the Church and would become a bishop. Entering the monastery at Ripoll as an adult in 1002, he jumped quickly from novice to abbot and embarked on a building program that transformed the churches Gerbert had worshipped in thirty-some years before. He introduced the Romanesque style, marked by tall square bell towers with many rows of arched windows, from Italy into Spain. The stone portal he commissioned for Ripoll is sixty-six feet wide, its arches and rows filled with hundreds of low-relief carvings telling stories from the Book of Kings and the Apocalypse. Ten larger figures, five on a side, portray the concept "harmony": On the left is an orchestra of viol, bells, recorder, and horn, the bishop conducting; on the right Christ gives the church to Peter, beside whom stand a bishop, a soldier, and a judge—a good emblem for the Catalonia Gerbert knew.

Oliba is also given credit for the huge jump in the number of books in the Ripoll library between 979 (when an inventory listed 65 manuscripts) and 1047 (when, just after his death, another inventory listed 246). Nor was it only the number of books that made the scriptorium famous: Many of the new books were about science.

One of the most controversial books still exists. Known as Ripoll 225, it is a collection of illustrated treatises on geometry and astronomy, including a description of how to use an astrolabe. Brief sections are literal translations from al-Khwarizmi's Arabic book on the astrolabe. Scholars have argued bitterly over whether Ripoll 225 is old enough for Gerbert to have seen it—or even to have written it himself. Today, based on the evidence of paleography—which examines how the parchment was prepared, the ink made, the letters shaped—the manuscript is thought to have been made in the eleventh century, during Oliba's time, at least thirty years after Gerbert left Catalonia. And yet it is not the translator's rough draft, with the cross-outs and additions and corrections a draft would have. It is a clean copy of something older, some translation of Arabic science that Gerbert might indeed have seen.

Another incident links Ripoll to Arabic science and both to Gerbert's circle of friends. Ripoll, too, had a new cathedral under construction while Gerbert was there. In 977, seven years after Gerbert left Spain, it was consecrated. Bishop Ato was dead. Garin of Cuxa was in Venice. But several of Gerbert's friends attended the ceremony. Miro Bonfill again wrote the consecration speech. In the audience were Count Borrell and a deacon of Barcelona, Seniofred, known by his nickname, Lobet; in 984, Gerbert would write to Lobet requesting "the book *De astrologia*, translated by you," which could be a treatise on the astrolabe or one on astrology, translated from Arabic.

Also attending the consecration ceremony was a monk named Vigila from the monastery of Albeda in the kingdom of Navarre. Vigila is famous for copying Isidore of Seville's encyclopedia, a project he is thought to have completed in 976. This copy, now known as the Codex Vigilanus, is the earliest Latin manuscript to contain what we call Arabic numerals—but which al-Khwarizmi, in the first book on this numerical system, called "Indian numerals." To Isidore's description of arithmetic, Vigila added a comment: "It should be noted that the Indians have an extremely subtle intelligence, and when it comes to arithmetic, geometry, and other such advanced disciplines, other ideas must make way for theirs. The best proof of this is the nine figures with which they represent each number no matter how high. This is how the figures look." Then he lists them, from 9 to 1, shaped a little differently than we would write them today.

Vigila does not sound as if he were announcing a great discovery, only fitting into Isidore's seventh-century text a bit of knowledge that had become common in the intervening three-hundred-some years. We don't know how or when he learned of Arabic numerals—nor whether he spoke of them with Miro or Lobet during his visit to Ripoll. But the Codex Vigilanus and Ripoll 225 prove that Arabic science and mathematics were making their way north from Islamic Spain around the year 1000, with the eager assistance of Catalan churchmen.

Miro Bonfill, the cousin of Count Borrell (and the count of Besalu in his own right), may have been Gerbert's closest friend in Spain. Miro became bishop of Girona in 971; in the same year—the year after Gerbert left Spain, if our dates are correct—he was sent on his own embassy to Cordoba. Miro left no account of his mission. But we can learn something about him and his knowledge of Arabic science from his other writings: the speeches celebrating the new churches built at Cuxa in 974 and Ripoll in 977, a charter dated 976, and a book on astrology, in which he wrote: "What follows now has been translated by the wisest scholar among the Arabs, as he was instructing me."

"The wisest scholar" could have been Maslama of Madrid, the chief mathematician and astronomer during al-Hakam's reign. His school was supported by the caliph, who provided books and other resources, such as astrolabes. Maslama produced a renowned star table, drawn up in 978, that adapted the work of al-Khwarizmi to the coordinates of Cordoba. He wrote a commentary on Ptolemy's *Planisphere* and a treatise on the astrolabe. He was also an astrologer: It was common for Arabic mathematicians and astronomers to tell fortunes. It was lucrative and (depending on the fortune) kept their patrons happy. Maslama lived until 1007.

Gerbert, likely, was also an astrologer. When Richer of Saint-Remy wrote that Gerbert "studied mathematics extensively and successfully" under Bishop Ato, he did not use the common Latin word *mathematica*. He used *mathesis*. We don't really know what Richer's idea of *mathesis* was. Boethius is the only other author known to use *mathesis* to mean mathematics. It more often meant astrology.

Miro may have been Gerbert's mentor in this science. His astrology book, now in the national library in Paris, is the oldest left from the Middle Ages. It shows an uncommon deftness with numbers: Its hundreds of multiplications and divisions contain absolutely no errors. The playful, provocative writing style, and the Catalan provenance of the manuscript, further link it to the bishop of Girona, though Miro did not sign this work. He did sign his speeches consecrating the churches at Cuxa and Ripoll, as well as the charter of 976. Like the astrology book, these writings are rife with puns.

Miro was obsessed by complexity. He regularly used synonyms and words so rare that his readers (or listeners) needed a glossary to understand them—and in fact, all of his odd word choices can be found in a set of glossaries in the library of Ripoll. These glossaries were, luckily for the history of science, in Barcelona being rebound when the library at Ripoll burned down in 1835. There are five in one volume, thirteen in another. Some offer synonyms for unusual Latin words found in Isidore of Seville's encyclopedia or in the classics of Roman poetry. Others are bilingual—Greek to Latin—or even trilingual—Hebrew to Greek to Latin. Miro played with them all.

Nor did his love of complications end with words: Miro played with numbers as well. Instead of "twenty-eight," he said "four times seven." Not "six months," but "twice three months." To say "976" in his charter of that year takes him three lines.

This puckish wit seems unlike the man Gerbert addresses in his letter, dated 984. Gerbert begins, "The great reputation of your name, indeed, moves me not only to see and speak with you, but also to comply with your orders." He would have done so sooner, he explains, but he had been engaged by Emperor Otto II; freed of his obligations by the emperor's death, "it is right for me both to talk with friends and to obey their commands." He closes with his request for the book *On the Multiplication and Division of Numbers* by Joseph the Wise. This extremely formal letter is the only one in Gerbert's collection addressed to Miro, and, sadly, Miro died before it could reach him. And yet it makes three things clear: Gerbert has corresponded with Miro before (or at least received orders), considers him a friend, and expects him to have a copy of the math book by Joseph the Wise.

And yet, a souvenir of their friendship—and Miro's playfulness—does remain. In the cathedral at Elne, on the Mediterranean coast close to the modern border of Spain and France, is a large gray stone. Carved on the far left of one edge is the name "Miro." On the right, "Gerbertus."

The Elne stone was discovered in the 1960s when the altar of the cathedral was moved. The altar itself was a slab of white marble, just like the altar at Cuxa on which are inscribed more than a hundred names dating back to the tenth century, including that of Oliba of Ripoll. There was a tradition of such sacred graffiti: It marked the making of a vow. At Elne, the marble altar was bare—the incised stone was found *under* the altar, and the names were not visible. The stone had apparently been moved from its original site and reused. Based on cuts in the corners of the stone, it might once have served as the lintel of a doorway, perhaps the lintel into the crypt where the sacred relics were kept when not on display—a plausible place to make public note of a vow.

Stranger is the fact that both names are puns.

Gerbert's signature is very fine, deeply carved, its edges still sharp after all these years. It is about three inches in height and breadth and shaped like a cross: GER (space) BER, with the T above the space and the US (written in Latin as VS), twined around each other, below.

MIRO is as long but not as high as GERBERTVS, and not as well carved, and it is *backwards: Miro* in Latin means to look in a mirror.

Gerbert couldn't find an equivalent Latin pun for his name: It is Germanic. So this twenty-year-old monk matched his friend's playfulness with a set of intensely intellectual puzzles. Pieced together, they reveal how he thought of himself as a young man—and how his sojourn in Spain had affected him.

To make his name into a cross, Gerbert pulled out T-V-S, recognizable to any medieval churchman as the Trinity, *Thevs Verbum Spiritvs*, or Father, Son, and Holy Ghost. The letters could also stand for *Tav Votvm Solvi*, "I have accomplished my vow to the Cross." He constructed the cross itself with one letter, two letters, and three plus three letters.

Graffiti carved by Gerbert at the cathedral of Elne. Like his friend Miro's signature, which is backwards, Gerbert's autograph is a pun, signifying church, wisdom, and empire.

According to the accepted mathematical theory, one, two, and three were the basis of God's creation of the universe, making number the key to wisdom. Finally, the letters were to be read left to right, then up to down, making the symbol known as the chrismon, still used in baptisms. The chrismon is the sign of the emperor Constantine, the first Christian emperor, and the one who moved the capital of his empire from Rome to Constantinople.

To read GERBERTVS was thus to summarize the three great forces acting on Gerbert's life: church, wisdom, and empire. The Church would be his home, the search for wisdom—through numbers—would be his passion, and the restoration of the empire, and its return to Rome, would be his lifelong goal.

CHAPTER IV

The Schoolmaster of Reims

In December 970, Gerbert left Spain, accompanying Count Borrell and Bishop Ato on a mission to Rome. Winter seems an odd time to attempt crossing the Alps, but there are many stories of similar expeditions. Odo of Cluny crossed in January. As the custom was, he hired local guides—Muslims, who from their fortress at Saint-Tropez on the coast had ruled the Alpine passes for almost a hundred years. Even then, the road was perilous. A blizzard struck at sunset. "We were so covered with snow and our limbs so frozen that we could not speak," one of Odo's companions wrote. "Suddenly, the horse on which our father was sitting slipped sideways and they both fell together down the steep slope. Letting go of the reins Odo raised both his hands to heaven as he fell, and immediately his arms found the branch of a tree from which he hung suspended until those in front turned back at his cries and rescued him. . . . The horse was never seen again."

River crossings could be chancy at any time of year. Odo was in midstream once when "one of the horses kicked at another and struck the side of the boat in a place where there was a knot in the planking. As soon as the side of the boat was pierced, so great a torrent of water came in through the hole that the boat was quickly filled." Odo reached shore only "by the manifest help of God."

Richer of Saint-Remy wrote of reaching a bridge on a rainy night at dark. It was "pierced with holes so large and so numerous that [we] . . .

would have had difficulty passing it even in the daytime." His escort, an experienced traveler, had a plan. "Over the gaping holes, he placed his shield for the horses to step on, or one of the loose boards that were lying around, and sometimes bending, sometimes straightening, sometimes on tiptoe, sometimes running, he succeeded in getting the horses and me across."

Then there were the tolls—or bribes. Count Gerald the Good of Aurillac, on his way to Rome, was stopped at Piacenza by the cleric in charge of the ford, who "for some reason . . . was in a very bad temper, flinging angry words about." Count Gerald "subdued" him with some "small gifts," and the man not only took them across the Po River but also refilled their flasks with wine. ("Small" should be read in the context of the count's well-known generosity: The Muslim guides who controlled the Alps "thought nothing more profitable than to carry Gerald's baggage through the pass of Mont Joux.")

Finally, the travelers had to beware of brigands. Count Gerald had just arrived at the city of Asti when a thief stole two of his packhorses. "Coming to a river, he was not able to get them across before he was taken by Count Gerald's men." Having retrieved his horses, the magnanimous count pardoned the man.

Count Borrell and Bishop Ato were attacked on their way home from Rome—but not by such simple thieves. They had gone to Rome to convince the pope to separate the churches of Catalonia from the archbishopric of Narbonne. The pope agreed. In a series of five papal bulls, he established a new archbishopric of Vic with Ato at its head. In revenge, the archbishop of Narbonne saw to it that his new rival never reached home: Ato was murdered in August 971.

Gerbert was not with him. He had impressed important people in Rome. According to Richer of Saint-Remy: "The pope did not fail to notice the youth's diligence and will to learn. And because *musica* and *astronomia* were completely ignored in Italy at that time, the pope through a legate promptly informed Otto, king of Germany and Italy, that a young man of such quality had arrived, one who had perfectly mastered *mathesis* and who was capable of teaching it effectively."

Otto the Great, who had been crowned Holy Roman Emperor in 962, took Gerbert into his court to tutor his heir, Otto II, then sixteen to Gerbert's twenty. Otto II was not a scholar by nature. Gerbert may have improved his Latin—he mentions Otto's "Socratic disputations" in a letter, and Otto later proved himself a book-lover—but Gerbert's mastery of the quadrivium was not appreciated until a year later, when Archbishop Adalbero of Reims came to visit the pope.

Adalbero, brother to the count of Verdun, wanted to improve the teaching of the seven liberal arts at his cathedral's school. Reims, now in the shadow of Paris, was then the leading city in France. Kings were anointed in the Reims cathedral, and the point of the cathedral school was not to turn out country priests but to train young noblemen for the king's service as bishops. It's a misconception that a bishop or archbishop had to be pious. They were as much counts and courtiers as churchmen, valued for their tact and managerial skills. And for this they needed the best education.

"While he was thinking along these lines," Richer of Saint-Remy wrote about Adalbero's plans for the school at Reims, "Gerbert was directed toward him by God himself." Or, as he enthused elsewhere, "when Divinity wished to illuminate Gaul, then shrouded in darkness, with a great light," it inspired Count Borrell to bring Gerbert to Rome, where he would meet the archbishop of Reims.

Gerbert needed a new post. After Otto II wed a Byzantine princess, Theophanu, on Easter Sunday 972, he no longer required a tutor. At the wedding, Gerbert struck up a friendship with Gerann, who taught the trivium of grammar, rhetoric, and logic at Reims. Gerann had sought out the young scholar, at Adalbero's request, to recruit him. Gerbert agreed to teach the quadrivium, the mathematical arts, at Reims in return for lessons from Gerann in dialectic. Taking his leave of the emperor, he rode with Gerann to Reims and became a canon of the cathedral. When Gerann died a few years later, Gerbert was named schoolmaster, a position he held, with only a short interruption, until Adalbero's death in 989, when Gerbert replaced him as archbishop.

Noble and wealthy and well aware that his diocese was the most powerful in France, Archbishop Adalbero of Reims was better called a prince-bishop, being the holder of many castles and overlord of several counts himself. His church lands stretched from the suburbs of Paris well across the border of France into the Holy Roman Empire, giving the archbishop ready access to the emperor, who likewise made sure the archbishop was his man.

To King Lothar of France—already under the eye of his mother Gerberga, sister to Emperor Otto the Great, and his uncle Bruno, the emperor's brother—Adalbero, when he was elected to the post in 969, seemed yet one more imperial spy. Eventually Lothar and his son, Louis V, would use a harsher term, *traitor*—an accusation they would extend to Gerbert, too, for Gerbert became Adalbero's secretary and confidant. To Gerbert, Adalbero was the pilot of the ship, the equipoise of the balance. "We were of one heart and soul," Gerbert wrote upon his friend's death in 989. Without him, "one might think the world is slipping into primordial chaos."

They were joined not only by their love of books and learning: Both were captivated by the idea of empire. The rightful order of the universe, they agreed, included an emperor, sovereign over all the Christian world, to whom the king of France would naturally pay homage. This respect for hierarchy in no way lessened King Lothar's honor, they argued. He was the representative of Christ within his realm. His dignity was reflected by the wealth and splendor of its churches.

The first thing Adalbero did upon becoming archbishop of Reims was to raze the vaulted foyer of the cathedral and make it "more worthy" of his king, according to Richer of Saint-Remy—even though that meant destroying the ramparts built to protect the church from Viking raids. He raised a bell tower. He commissioned a cross of gold for the main altar and surrounded it with balustrades sparkling with precious stones. He gave the church elegant new reliquaries, a seven-branched candelabrum, a portable altar with gold and silver statues of the four evangelists, marble floors, brilliant frescos on the walls, and some of the earliest stained-glass windows, or, as Richer described them, "windows containing various stories." Finally, he hung golden crowns over the altars, symbolically linking earthly kingship to the eternal throne of Christ the Lord. He hoped Lothar would rise to the occasion.

Adalbero brought order to the many monasteries under his care and cracked down on the independent canons of Reims, building a dormitory where they were required to sleep. They had to eat together in silence, take part in the night offices, and wear identical, sober clothing. We don't know what the canons had been wearing before, but the monks like Richer had to make do without their "bonnets with long ear flaps," their "excessive breeches whose leggings stretch the length of six feet and yet do not protect . . . the shameful parts from onlookers," their costly tunics that were "so tight across the buttocks" that the monks' "asses resembled those of prostitutes," and their tall, tight boots with up-curved toes.

Yet the rules at a cathedral were looser than those of a monastery—it was more like a gentlemen's club than a cloister. The canons of a cathedral sang Mass and celebrated holy festivals. They maintained the church buildings, the altars and ornaments, relics and vestments, and oversaw the church lands and other revenues, such as the taxes on markets, the tolls on bridges and roads, the fines from lawsuits, the collecting of the tithe, and the coining of money. They cared for the poor and sick, gave hospitality to travelers, and acted as go-betweens in feuds, often being the ones to ransom captives.

Being a canon, not just a monk, was a mark of status. Canons could eat meat, wear linen, and—most important to Gerbert—acquire possessions of their own. While monks were only allowed to use the word "my" to refer to their parents or their sins, canons owned books, rings, coins, goblets, rugs, barrels of wine, Saracen slaves (according to the wills of some Spanish canons), houses, and land. Many canons came from wealthy families who had dedicated a son to the church, not only to ensure the family's salvation, but also to have influence with the archbishop. With the boy, or oblate, had come an estate: Its income was meant to support the child, in style, throughout his life, and it was the oblate, once grown, who had control of it.

Gerbert's position as master of the cathedral school likewise brought him wealth. In his letters he mentions the houses that "at great expense we built, together with their furnishing. Also the churches that we obtained by solemn and legitimate gifts," and the vast sums he was spending on books. When he left Reims in 981 to become (briefly) the abbot of Bobbio, he scandalized the Italian monks with the number of servants and quantity of goods he brought from the north. His enemies whispered to the emperor that he must secretly be keeping a wife to require such a lavish household. After he escaped Bobbio in 983 and fled empty-handed back to Reims, he bemoaned for years the fact that "the best part of my household paraphernalia" had been left behind in Italy. He had been fond of his treasures.

As long as a canon made it to church for prime at sunrise and vespers at sunset, he could be "in the world," not cloistered like a monk, for most of the day. At some cathedrals, canons spent much of their time managing their property: their vineyards, estates, and townhouses. They were repeatedly censured for disgraceful behavior, including gambling, hunting, and the keeping of concubines. Other cathedrals were more like universities, with most of the day given over to classes. The first universities, founded in the 1200s, were cathedral schools that had cut their ties to the Church.

Reims, through the combined efforts of Adalbero and Gerbert, developed into a proto-university. Gerbert taught all seven of the liberal arts after Gerann's death, being expert in the trivium as well as the more advanced quadrivium. Students flocked to his school from throughout France and Germany; they even crossed the Alps from Italy. Among them were sons of noblemen, being readied for court life or positions high in the Church. Between 972, when he first came to Reims, and 996, when he left in disgrace, Gerbert taught, for example, thirteen future bishops or archbishops, six abbots of important monasteries, Emperor Otto III's chancellor, the secretary to Emperor Henry II, the future Pope Gregory VI, and King Hugh Capet's son Robert the Pious, who would rule France from 996 to 1031. Says *The Life of King Robert,*

"His mother sent him to the school of Reims and confided him to master Gerbert to be taught by him and instructed in the liberal arts in a manner in every way pleasing, by his virtues, to God."

Not all of his students were of noble blood. Many were simply inquisitive monks and canons, wandering scholars who in the course of their education "pillaged many schools," in one medieval description, or visited "masters of schools far away, like a prudent bee which goes from flower to flower collecting sweeter honey." To attend, they needed permission from their abbots or bishops, and Gerbert and Adalbero did their best to arrange it. Adalbero writes to the abbot of Ghent, "We have adopted one of your brothers, but you are detaining one of ours who ought to return." To the archbishop of Trier, Gerbert says, "If you are wondering whether you should direct students to us . . . on this matter we have an open mind."

Gerbert's letters hint at his teaching philosophy. He speaks of the importance of "a mind conscious of itself," of studying mathematics "for the utmost exercise of the mind" and astronomy "in order not to grow inwardly lazy." To the abbot of Tours, he explains the importance of rhetoric to those who are "busied in affairs of state": "For speaking effectively to persuade and restraining the minds of angry persons from violence by smooth speech are both of the greatest usefulness. For this activity, which must be prepared beforehand, I am diligently forming a library."

And though learning required books—ever more books—teaching did not. Gerbert was an orator, not a writer. He shared what he knew through speech and demonstrations, not texts. Only very reluctantly, under pressure of a student's request, did he write things down. For instance, when Gerbert wrote his *Book on the Abacus*, he sent it off with a letter that began: "Only the compulsion of friendship reduces the nearly impossible to the possible. Otherwise, how could we strive to explain the rules of the abacus unless urged by you, O Constantine, sweet solace of my labors?"

Constantine, who seems to have been Gerbert's favorite student, was raised in the monastery at Fleury, near the French city of Orleans. Like

Gerbert, he was singled out for his intelligence and sent off to learn the quadrivium. Of aristocratic birth and said to be devastatingly beautiful, he attracted a following at Reims. When the time came for him to leave, an unknown monk (probably not Gerbert) wrote a parting ode dripping with learned allusions, Greek words, and strained grammar. Wisdom herself was Constantine's teacher. The goddess had built a temple in this "man magnificent above others and always loveable." That temple shines with "the excellent light of virtues," with "nobility of merits" and "probity of manners." The poet's beloved Constantine is chaste as a dove, cunning as a serpent, a mirror of justice, the light of the learned. "These pages will bring happiness when your presence cannot," says the poet of his "sweetest" friend.

Constantine and Gerbert kept up a correspondence—dense and technical and including many of Gerbert's known scientific writings—on such subjects as how to make a hemisphere for studying the heavens, the solution to a seemingly insolvable problem on sesquiquartal numbers in Boethius's *On Arithmetic*, the equally abstruse theory of superparticular numbers from Boethius's *On Music*, and the rules of the abacus. Constantine copied and shared these scientific notes from his teacher with other like-minded monks.

Constantine also preserved Gerbert's formal letter collection. When Gerbert was forced to leave Reims, he sorted through his letters and confided a selection to Constantine. Shortly after Gerbert was named pope in 999, Constantine made a copy of these papers. Though the originals are lost, Constantine's copy still exists in the library of the University of Leiden. One page gives the pen name of the scribe, *Stabilis*, meaning "stability" or "constancy."

Gerbert's letter collection was carefully edited. Like Cicero, Gerbert chose letters that showed his rhetorical skill and historical importance. In some ways, he was creating a textbook on rhetoric; notably, he chose not to include any of his scientific papers. He was also writing his autobiography. His letters reveal what was important to him: Second only to lifelong learning was friendship, and the two were inextricably intertwined.

Constantine made this copy of Gerbert's letter collection sometime between 999, when Gerbert became pope, and 1014, when Constantine died. Just above the large capital letter, which falls in the middle of a page, you can read: *Incipit exemplar epistolarum Girbirti Papa . . .*, "Here begins the copy of the letters of Pope Gerbert . . ."

"I do not know that divinity has given to mortals anything better than friends," Gerbert wrote.

Friendship is honorable and sacred, he told the abbot of Tours. "Since you hold the constant memory of me among things worthy of honor, as I have heard from a great many messengers, and since you bear me great friendship because of our relationship, I think I shall be blest by virtue of your good opinion of me, if only I am the sort of man who, in the judgment of so great a man, is found worthy to be loved."

This conflation of love and friendship is found throughout Gerbert's letters. Writing to a monk from Aurillac, Gerbert describes a teaching tool he had devised "out of love" for his students: It was a "table of the rhetorical art, arranged on twenty-six sheets of parchment fastened together in the shape of an oblong . . . a work truly wonderful for the ignorant and useful for the studious in comprehending the fleeting and very obscure materials of the rhetoricians and keeping them in mind." He closes, saying, "Farewell, sweetest brother; always enjoy my love which is equal to thine, and consider my goods to be for both of us."

Many of his friends received such fond treatment. To one, he writes, "I am unable, sweetest brother, to display my genuine affection for you." To others, "We congratulate you, sweetest brother, . . ." "your good will, beloved brother, . . ." "your request is indeed a large order, dearest brother, . . ." "I am well aware that you understand the emotions of my mind, and on this account I love and embrace you, . . ." "your request, sweetest brother, so often repeated. . . ."

All of this "sweetest brother" stuff, like his epithet for Constantine, "sweet solace of my labors," sounds a little precious, even pathetic. But we should not be fooled into thinking the love Gerbert pined for was homoerotic. Though his sweet words were uttered monk to monk, they were also uttered king to king. The language of love was the speech of courtiers. Here is the French king, writing to the Byzantine emperors—through Gerbert's pen—asking for a royal bride for his son: "Not only the nobility of your race but also the glory of your great deeds urges and compels us to love you. You seem, indeed, to be such preeminent persons that nothing in human affairs can be valued more highly than your friendship."

Love was a sign of the highest respect. One loves, as the king's letter outlines, what is noble and glorious in another person. Fondness was possible only between two virtuous souls, for true friendship, as Cicero said, was the love of goodness in another. Gerbert's school at Reims was founded on this Ciceronean code of friendship, on the mutual desire of friends to better each other. It was for love of his students in this classical way that Gerbert "expended quantities of sweat," as Richer of Saint-Remy put it, on his teaching.

Yet friendship was also useful, Gerbert noted to the abbot of Tours. "Because I am not the sort of man who, with Panetius, sometimes separates the honorable from the useful, but rather with Cicero would add the former to everything useful, so I wish that this most honorable and sacred friendship may not be without its usefulness to both parties." How could the abbot best demonstrate his friendship? Gerbert attached a list of books he could have copied and sent to Reims.

Given his own reticence, our best window into Gerbert's school is the description by Richer of Saint-Remy in his *History of France*, written between 991 and 997. This was twenty years after Gerbert had come to Reims and coincides with a time when his political troubles were at their height.

Richer begins by outlining Gerbert's teaching of the trivium. To learn Latin grammar, his students studied Cicero; the poets Virgil, Statius, and Terence; the satirists Juvenal, Persius, and Horace; and Lucian the historiographer. "Once his pupils were familiar with these and acquainted with their style, he led them on to rhetoric," writes Richer. "After they were instructed in this art, he brought up a sophist on whom they tried out their disputations, so that practiced in this art they might seem to argue artlessly, which he deemed the height of oratory." Concluding the study of dialectics, Gerbert read aloud from "a series of books"—most of them by Boethius—"accompanied with learned words of explanation."

Moving on to the quadrivium, Richer depicts Gerbert as a master of visual aids. "He demonstrated the form of the world by a plain wooden sphere, thus expressing a very big thing by a little model," according to Richer. "This is how he produced knowledge in his pupils." He made celestial spheres for observing the stars, to explain the motion of the planets, and for learning the major constellations. He had a shieldmaker construct an abacus: "Its length was divided into twenty-seven parts, on which he arranged nine signs expressing all the numbers." Using a thousand counters made of horn and marked with these "nine signs," he could multiply and divide with such speed that "one could get the answer quicker than he could express it in words." To teach music theory, Gerbert used a monochord, a simple one-stringed instrument.

Richer does not go into much detail. In many cases his descriptions of Gerbert's methods are unclear—he seems not to have been mathematically inclined himself. About the abacus, he says, "Those who wish to understand fully this method should read the book which he wrote to the scholasticus Constantine, where one will find this subject fully treated." While discussing Gerbert's celestial spheres, he simply breaks off: "It would take too long to tell here how he proceeded further; this would sidetrack us from our subject."

Nor can his account be entirely trusted. History for Richer was a literary art: He saw nothing wrong with putting into the mouth of Charles of Lorraine, who fought Hugh Capet for the French throne from 987 to 991, a speech by King Herod from the fourth-century Latin translation of Flavius Josephus's *The Jewish War*. Another of Charles's moving speeches comes directly from Sallust.

Moreover, Richer fiddled with facts. He took episodes from the annals of Flodoard of Reims, who died in 966, and changed such things as the size of an army, the number of casualties, the locations of battles, and who won and who lost. He saw the changes as improvements: "I think that I have done well enough by the reader," he wrote, "if I have arranged all things credibly, clearly, and briefly." We should not expect him to be accurate or complete. His definition of "history" was not the same as ours.

Richer knew Gerbert, but he was not Gerbert's student or admirer, as some historians have claimed. He was a monk at the monastery of Saint-Remy a few miles outside of Reims and the same age or older than Gerbert. The two had serious political disagreements, Richer being a partisan of the last Carolingian, Charles of Lorraine, while Gerbert, as we will see, was central in placing the challenger Hugh Capet on the French throne. Richer had reasons to flatter Gerbert, and reasons to distance himself. While Richer was writing his history, Gerbert, threatened with excommunication by the pope, was struggling to hold onto his position as archbishop of Reims. There are hints that Gerbert may have commissioned Richer to write about him in hopes of salvaging his reputation: Some passages in the *History* seem to have been cribbed from Gerbert's own letters. Yet, as he revised his book—and as it became clear that Gerbert would be evicted from his post at Reims— Richer put a subtle twist on events, calling Gerbert's actions and character into question.

When Richer died, with the history unfinished, Gerbert apparently got hold of the only copy—and hid it away. It did not circulate in the

Middle Ages. No copies were made. All we have is Richer's very messy rough draft. It was discovered in the 1830s among the books of Gerbert's last student, Emperor Otto III, in the library of the cathedral of Bamberg. With its scribbles and cross-outs, tipped-in pages and marginal notes, asterisks and erasures and several colors of ink, the manuscript is evidence of a complicated writing process, and a writer trying to make up his mind.

As a close reading of the manuscript shows, Richer wrote the story of Gerbert's school separately and struggled to add it to his work-in-progress. He erased and rewrote the text that preceded it to make a better transition, but the section remains jarringly different from what surrounds it. And while the entire *History* is dedicated to Gerbert, the dedication seems contingent on Gerbert's keeping the archbishopric. Although the king of France had appointed him to the position, the pope refused to consecrate him, arguing that another candidate had a better claim. Describing that seven-year-long dispute between the king and the pope, Richer sides with Gerbert's enemies and subtly contradicts the official record that Gerbert wrote. Gerbert comes off so poorly that it seems Richer never meant for him to read the final, much-revised account.

Some of what Richer says about Gerbert can be corroborated. We know Borrell and Ato went to Rome—two of the five papal bulls still exist in Vic—and the cathedral records show Ato died before reaching home. From Gerbert's letters we know he met the emperor, briefly taught his heir, and then went to Reims to teach. Gerbert wrote about the abacus, the celestial spheres, and some other visual aids, but his descriptions are vague: They assume that his correspondent has seen the object under discussion.

Since he did not begin keeping copies of his correspondence until ten years after he had left Spain, it's hard to say what Gerbert taught when he first arrived at Reims. What did Gerbert know of *musica, astronomia,* and *mathesis* that had so impressed the pope, the emperor, and the archbishop in Rome? What science had he learned in Spain? Gerbert left no scientific manuscripts, firmly dated before 970, to prove he learned anything extraordinary at all.

Yet Gerbert's Catalan friends were well placed to learn more, as translations were made from Arabic and new scientific instruments—and the knowledge needed to make them—seeped north. And Gerbert's letters prove they kept in touch. Whether he learned Islamic science in Catalonia between the years 967 and 970 or he learned it later—by correspondence course, as it were—we can see from the excitement his school aroused; by the new directions taken by Gerbert's students, and by *their* students; and by the criticisms of Gerbert's peers that the *mathesis*, *astronomia*, and *musica* Gerbert taught at Reims were unlike anything the Christian West had seen before.

PART TWO
GERBERT
THE SCIENTIST

I shall, if life continues, explain these matters to you more clearly, as much as is necessary for you to attain the fullest understanding.

GERBERT OF AURILLAC, C. 979

CHAPTER V

The Abacus

The rumors began seventy years after his death. Over time, they grew more and more surreal. By the twelfth century, a courtier could ask, "Who has not heard of the fantastic illusion of the notorious Gerbert?" He sacrificed to demons and summoned up the devil himself. He was a wizard, a necromancer. Through a ravishing witch, or a golden head, or a magical book he had stolen in Cordoba, or simply through "the stars," he foretold the future. He grew fabulously rich. He obtained everything he desired.

Ironically, both the honors Gerbert enjoyed in his lifetime and the contumely heaped on him later were due to one thing: what he taught at Reims. But Gerbert's genius is hard to pin down. His science has to be inferred, and the evidence is scanty. We have a few letters, the account in Richer of Saint-Remy's *History of France*, the fresh approaches of his students, and a handful of artifacts. One such artifact came to light in 2001: an actual copy of Gerbert's abacus board. It makes clear that when Richer described Gerbert's abacus, with its counters marked with "nine signs," he was, in fact, recording the introduction of Arabic numerals to France.

The abacus found in 2001 is a stiff poster-sized sheet of parchment; it was trimmed down and reused as a pastedown in the binding of the Giant Bible made for the abbot of Echternach sometime between 1051 and 1081 (see Plate 5). The Bible is owned by the national library of

Luxembourg, which had unbound it in 1940 in order to photograph the pages. The abacus sheet was taken out of the binding and stored in a box, where it remained, unrecognized, for sixty-one years.

Shortly after this find was announced, a librarian at the state archives of Trier identified a matching copy, smaller, but written in the same handwriting; it also came from the scriptorium at Echternach. This second abacus was bound into a very interesting manuscript. It could be the notebook of one of Gerbert's students. Following the abacus is a mnemonic poem on the names of the nine Arabic numerals and zero. The manuscript holds miscellaneous notes on multiplication and division, the use of Roman fractions, and the etymology of the word *digit*. These notes contain many corrections and erasures. They do not match any known source but seem to be the messy jottings of a single scholar, penned over a number of months or years. This student copied Gerbert's poem on Boethius into his notebook, along with other texts that reflect Gerbert's curriculum at Reims, as Richer describes it.

We can even guess at the student's name. The manuscript can be dated to 993 by its similarity to other large-format books made by a scribe historians have named "Hand B." One of these books is dedicated to the monastery of Echternach by an English monk named Leofsin. Leofsin moved to Echternach in 993. Before that, he had lived at Mettlach, alongside one of Gerbert's favorite students, a monk named Gausbert.

Gausbert is mentioned in several of Gerbert's letters. Some of these are addressed to Abbot Nithard of Mettlach, who himself had been Gerbert's student in the early 970s. Nithard's relationship with his former teacher was intimate—and a little prickly. "You think you alone bear burdens but you do not know what the overwhelming trials of others are," Gerbert wrote him in 986. At issue between the two men was a "treasure" belonging to Nithard that Gerbert refused to bring or send back to Mettlach. "Since men are tossed about by an uncertain fate, . . . why do you lay up a treasure for a bad turn of fortune by leaving it with me for so long a time? And, inasmuch as I, a trustworthy man, am addressing a trustworthy man, make haste. For, either the imperial court

will summon me quickly, or, more quickly, Spain, which has been neglected for a long time, will seek me again."

The treasure might have been a book or an abacus—or a monk.
Nine months earlier Gerbert had complained to Nithard: "Suddenly,
with no consideration for the shortness of the time, you force Brother
Gausbert to return with everything belonging to him. . . . You have said
that he is unwilling to return to the tedium of the monastery. If this is
so, how are you going to hold him after he has been returned?"

Gausbert is also mentioned in two letters to Nithard's superior, the
archbishop of Trier. "We have never tried to hold the monk Gausbert
against your wishes," Gerbert claims, though it is clear he is sorry to
see him go. "We ask only this of your customary good will, that you
exhibit kindness to him on account of our recommendation, and . . .
let him not lack the studies to which he has arranged to devote fuller
attention."

It's tempting to think Gausbert, bored by the tedium of the
monastery and worthy of continuing his studies, was the messy scholar
who brought his notes from Gerbert's classes to Mettlach. There, Gausbert shared them—including the abacus and the poem on Arabic numerals—with Leofsin, who carried them on to Echternach and,
ultimately, to us. For until the two abacus sheets made at Echternach
were identified, scholars did not agree that Gerbert had used Arabic
numerals or that his abacus was really anything out of the ordinary.
With copies of Gerbert's abacus now in hand, every history of mathematics will have to be revised.

Sometimes called the first calculator or even the first computer, an abacus can take many forms. The ancient Chinese developed the version
that springs first to mind: colored beads strung on wires set in a vertical or slanted frame. Gerbert's abacus did not look like this. Nor was it
like the ancient Roman abacus: a palm-sized rectangle of bronze or clay
with seven vertical grooves divided in half by one horizontal line. The
Roman abacus used balls, not beads, to represent numbers. A ball
parked in a groove below the line stood for one—one unit, one ten,

one hundred, and so on, up to one million. A ball placed above the line meant five of the same.

In Latin *abacus* means "table"; it may refer to a sideboard, a game-board, or a counting board like Gerbert's abacus, which was a simple grid of twenty-seven columns, scored or painted on a flat surface. Later, merchants and bankers found having a counting board so useful that they drew them on tabletops, which they called "counters": That is why we now do business "over the counter."

Gerbert's abacus board introduced the place-value method of calculating that we still use today. Each column on Gerbert's abacus represented a power of ten. The "ones" column was placed farthest to the right, and the numbers increased by a multiple of ten, just as we read numbers now, in each column to the left. The twenty-seven columns were grouped in threes, linked by swooping arches, just where we would divide a large number by commas. Each group of three was labeled with an Arabic numeral, from 1 to 9. With twenty-seven columns, Gerbert could add, subtract, multiply, or divide an octillion (10^{27}). There was no practical reason for octillions—Gerbert was merely showing off. Witness the air of braggadoccio with which he writes to the emperor in later years, "May the last number of the abacus be the length of your life." That's 999,999,999,999,999,999,999,999,999. Don't even think of writing this in Roman numerals.

But the twenty-seven-column counting board caught on. A monk named Bernelin wrote a *Book of the Abacus* while Gerbert was pope. Protesting that it was presumptuous of him to try to better Pope Gerbert's brief and subtle work, he nevertheless designed an abacus with thirty columns (Gerbert's twenty-seven plus three for fractions). He suggested it should be drawn on a polished table.

Gerbert had his counting board constructed by a shieldmaker, according to Richer of Saint-Remy. To make a shield, a piece of prepared skin is stretched over a large wooden frame. A shieldmaker would know how best to get a wide, smooth surface that could be painted on. Such a counting board would be light, sturdy, and more portable than a tabletop. For a teacher, it would make a fine visual aid.

A shieldmaker also had the tools to cut a thousand *apices*, or "counters," out of cow's horn. These counters (some modern translators use the term "markers" to avoid confusion with the counting board itself) looked rather like checkers, with one important difference: Each was marked with an Arabic numeral, from 1 to 9. To calculate, Gerbert placed the counters on the counting board and shuffled them around. The speed with which he did so, said Richer, was astonishing.

To reconstruct the process takes some imagination. Gerbert's own *Book of the Abacus* provides little help. Answering the request of his friend Constantine of Fleury, Gerbert claimed it was "nearly impossible" to explain the rules of the abacus in writing and, moreover, he was out of practice: "Since it has now been some years since we have had either a book or any practice in this sort of thing, we can offer you only certain rules repeated from memory." Apparently he had been criticized for teaching this new math at Reims and saw the need to justify himself, at least to sympathetic ears, for he continued:

Do not let any half-educated philosopher think [the rules of the abacus] are contrary to any of the arts or to philosophy. For who can say which are digits, which are articles, which the lesser numbers of divisors, if he disdains sitting at the feet of the ancients? Though really still a learner along with me, he pretends that only he has knowledge of it, as Horace says. How can the same number be considered in one case simple, in another composite, now a digit, now an article?

Here in this letter, diligent researcher, you now have the rational method, briefly expressed in words, 'tis true, but extensive in meaning, for the multiplication and division of the columns [of the abacus] with actual numbers resulting from measurements determined by the inclination and erection of the geometrical radius, as well as for comparing with true fidelity the theoretical and actual measurement of the sky and of the earth.

The letter itself is hard to interpret. The person Gerbert calls a "half-educated philosopher" can possibly be identified, as we will see. The "actual numbers resulting from measurements" are also tantalizing. They could refer to measurements taken with an astrolabe or other scientific instrument, while the idea of "comparing with true fidelity the theoretical and actual measurement of the sky and of the earth" is a foretaste of the scientific method. Those who believe there was no experimental science in the Dark Ages, only memorization and appeals to authority, have never read the letters of Gerbert.

But the rules for the abacus that Gerbert appends to this letter provide no details on what it looked like, and very little help on how to use it. Constantine must have already seen an abacus board, for Gerbert does not explain how to make one. Instead he merely lists which column the result should be placed in if one multiplies a unit by a ten, a ten by a ten, a ten by a hundred, a hundred by a hundred . . . on up to a million by ten million. The rules for division are equally boring. He omits the rules for addition and subtraction, those being too elementary.

Gerbert does give some explanation of the twenty-seven columns. These were a sticking point for many learners, for whom the place-value system of arithmetic was radically new. What did the columns *mean?* Gerbert refers to them as "intervals," alluding to Boethius's use of the word in music theory. An interval was the distance between a low-pitched note and a higher-pitched note: the space between the two points. Elsewhere, Gerbert calls the columns "the seats of correct figures," analogous to Cicero's idea that topics are "the seats of argument." When preparing to debate, students were taught to organize their stock phrases, allusions, and other rhetorical flourishes in the rooms of their "house of memory." When preparing to calculate, they first arranged their numbers in their proper spaces on the abacus board. Three hundred sixty-five would be converted into a 3 counter placed in the hundreds column, a 6 counter in the tens column, and a 5 counter in the ones column. The same three counters, placed in different columns, could make 536 or 653. This was the key to the place-value system: The *place* where the counter sat determined the *value* of the number written on it, whether it meant five or fifty or five hundred.

The copy of Gerbert's abacus board in his student Gausbert's notebook is followed by a mnemonic poem on the names of the Arabic numerals. At the end of each line, the numeral itself is given. Significantly, the poem includes a symbol for zero. Some of the nine numbers, as shown here, are recognizable to modern eyes (if sideways or upside down), while others look very different.

A zero counter was not strictly necessary—to make 10, they just put a 1 counter in the tens column and left the ones column blank. For 100, the 1 counter was simply placed one column further to the left. But larger numbers, such as 10,001, might be confusing; the student's eye might not automatically connect the two 1 counters as parts of the same, composite number across so many blank columns. The mnemonic poem in the Trier manuscript includes a zero, looking like a spoked wheel, which Gerbert thought of as a placeholder. It filled the empty space, showing the column was in use. The idea that zero was an actual *number* would not arrive until much later. Ralph of Laon, who wrote an abacus treatise in about 1110, vaguely explains the zero by saying, "Even though it signifies no number, it has its uses."

Gerbert did not invent the counting board. Before his time people drew a grid on a flat surface and calculated with *calculi*—that is, pebbles. The word "calculate" originally meant nothing more than "move pebbles around." To add nine plus eight, you put a pile of nine pebbles in the units column. You put a pile of eight pebbles beneath it, in the same column. You smushed the two piles together and picked out ten pebbles.

Discarding nine of them, you put one pebble in the tens column (or two pebbles in the fives column, depending on how your abacus was set up). Then you counted what was left in the units column—seven pebbles—and wrote the answer in Roman numerals: XVII. For small numbers, this system was not very useful (you could do it faster in your head). For large ones, it wasn't convenient—you needed a huge bag of pebbles. To calculate tithes and taxes, a monk would more likely calculate in his head, using "finger numbers" to record the intermediate stages in a sum.

Finger numbers were not what we dismiss as "counting on your fingers." Known from ancient Greece to the Renaissance and in every culture from Europe to the Orient, the system called for extraordinary flexibility, with each joint of each finger moving independently—impossible for anyone with arthritis. Martianus Capella, a fifth-century scholar, complained that numbers above 9,000 called for "the gesticulations of dancers." There was also a great deal to memorize. The left hand was used to express numbers up to 99 (or XCIX). The right hand took care of the hundreds and thousands. Both hands together could account for any number up to 10,000 (MMMMMMMMMM).

The Venerable Bede, an Anglo-Saxon monk in Northumbria, England, describes finger counting enthusiastically in his book *On the Reckoning of Time*, written in about 725:

> When you say one, bend the left little finger and touch the middle line of the palm with it. When you say two, bend the third finger to the same place. When you say three, bend the middle finger in the same way. When you say four, raise the little finger. When you say five, raise the third finger. When you say six, raise the middle finger and bend the third finger down to the middle of the palm. When you say seven, touch the base of the palm with the little finger and hold up all the other fingers. When you say eight, bend the third finger in the same way. When you say nine, bend the shameless finger in the same way.

Yes, the shameless finger is the middle finger—still shameless today.

The numbers ten through ninety were also made with the left hand. Some were extremely complicated. For sixty, for example, first bend your thumb toward your palm "with its upper end lowered," like the capital letter for the Greek *gamma*. Then, keeping your thumb bent, place your forefinger over it, just below the thumbnail.

Similar contortions of the right hand made the hundreds and thousands. You could count higher by placing the left hand in various ways on the chest, back, or thigh. For 90,000, for example, "place the left hand on the small of the back, with the thumb pointing toward the genitals." For a million, join both hands with your fingers interlaced.

In addition to calculating, finger numbers were used as a secret code. Monks assigned a number to each letter of the Latin alphabet. By signaling 3–1–20–19–5–1–7–5, Bede reported, you could tell your companion, *Caute age*, "Be careful!" Medieval Arabic poetry, meanwhile, gives useful examples of how you could insult someone by displaying certain numbers: 93 meant the person was stingy, 30 meant he was picking lice off his private parts, 90 meant "asshole."

Gerbert was a finger-counter. Explaining the place-value system to Constantine he had said: "How can the same number be considered . . . now a digit, now an article?" Literally, the Latin *digitus* is "finger," and *articulus* is "joint." Gerbert was relying on Constantine's familiarity with finger numbers to give *digitus* a figurative meaning: A digit was any number placed in the first column of the abacus board, what we still call the digits column. For this reason, Gerbert's abacus is said to be the first calculating device to function *digitally*.

Gerbert would also have been very familiar with Roman numerals, which themselves recalled a basic method of finger counting. One (I) is a single finger. Five (V) is an open palm (think of the thumb as one arm of the V and the four fingers as the other). Ten (X) is two open palms. These three symbols are older than the alphabet; the higher numbers (L for 50, C for 100, and M for 1,000) originally came from Greek letters.

Medieval Greek, Arabic, and Jewish cultures generally used letters to stand for numbers—the first nine letters of the alphabet represented the first nine numbers. Bernelin, describing his thirty-column abacus around the year 1000, says the counters can be marked either with the nine Arabic numerals or with the first nine Greek letters, as if this number system might be more familiar to his readers. Scholars have long used Bernelin to argue that Gerbert used Greek letters—and didn't know Arabic numerals at all, much less introduce them to Christian Europe.

Yet the two copies of Gerbert's abacus board made at Echternach in about 993 prove that he not only taught math with Arabic numerals but was also the first Christian in the West known to do so. Altogether, eight manuscripts contain abacus boards based on Gerbert's. One was created at the monastery of Fleury, where Gerbert's student Constantine lived; it now resides in a library in Bern, Switzerland. Incorporated into the design, in the V-shaped gaps between the swooping arches, is a line of Latin hexameter verse: "Gerbert gave the Latin world the numbers and the figures of the abacus." The verse also appears, this time just beneath the Arabic numerals, on an abacus board now in the Vatican.

Many of these abacus boards give the names of the numbers. From 1 to 9 they are: *igin, andras, ormis, arbas, quimas, calctis* or *caletis, zenis, temenias, celentis,* and for zero, *sipos* or *rota*. The same names appear in the poem in Gausbert's notebook and, as we will see, in a poem Gerbert himself wrote while he was abbot of Bobbio in 983. Where these names came from is not known. *Arbas, quimas,* and *temenias* (4, 5, and 8) are likely distortions of Arabic words, while *igin* sounds as if it were Berber. The others may be Greek, Hebrew, or Chaldean.

Precisely when the concept of expressing all numbers using only nine symbols and a zero came west from India is also unknown. In Arabic, the symbols were sometimes called *ghubar* numbers, or "dust" numbers, because they were easy to write on a board lightly covered with sand. Dust numbers were known in Syria by 662, when Severus Sebokt wrote: "I will omit all discussion of the science of the Hindus, . . . discoveries that are more ingenious than those of the Greeks and the Babylonians; their valuable methods of calculation; and their computing that surpasses description. I wish only to say that this computation is done by means of nine signs."

The system was still new in ninth-century Baghdad, when al-Khwarizmi wrote *On Indian Calculation*. Al-Khwarizmi sought, in his words, to "reveal the numbering of the Indians by means of nine symbols." His book was brought to Spain, but the new numbers didn't immediately catch on with merchants or administrators, who preferred to use letters of the alphabet to stand for numbers. Lest we think the Spanish were overly conservative, these numbers were not used by Baghdad's businessmen either. Abul Wafa al-Buzjani is known to historians of mathematics for his advances in trigonometry: He devised the tangent function and found a way to calculate sine tables accurate to eight decimal places. He used Arabic numerals for his theoretical work, but not in his *Book of Arithmetic Needed by Scribes and Merchants*, written in Baghdad between 961 and 976. Here, he teaches how to calculate with whole numbers (including negative numbers) and fractions, how to find the volume of a solid body, and how to measure distances. He discusses taxes, exchange rates, maintaining an army, and building dams, all while using letters of the Arabic alphabet to stand for numbers.

The oldest known Latin manuscript to contain Arabic numerals is the Spanish monk Vigila's copy of Isidore of Seville's encyclopedia, which echoes al-Khwarizmi's description of them as Indian numerals. It is dated 976, six years after Gerbert left Spain. The *Book on the Abacus* Gerbert wrote for Constantine—and the letter in which he complained, it "has now been some years since we have had either a book or any practice in this sort of thing"—cannot be dated precisely. Because he refers to himself as *scholasticus* ("schoolmaster" or "teacher"), we can assume he wrote it at Reims between 972 and 980, when these ideas were still very new.

Gerbert's *Book on the Abacus* was very popular: It still exists in thirty-five manuscripts. Over the next 150 years, it inspired fifteen other monks to write their own treatises on the abacus. Historians call them "dry" and "dull" and are puzzled when the monks themselves describe their abacus studies as "arduous" and "most difficult," easy to understand on the surface, but hard to those who looked deeper. In the early 1100s, William of Malmesbury mocked his more mathematical peers, saying the rules of the abacus were "barely understood by the perspiring abacists themselves." Yet at the same time, Ralph of Laon wrote about the abacus: "From Gerbert, a man of the highest prudence, whose very name is wisdom, the channel of this science has run down to our own times, even though it is a narrow one." The Laon school can be traced to Gerbert through its influential bishop, Ascelin (a nephew of Archbishop Adalbero of Reims), who had studied with Gerbert and outlived his mentor by nearly thirty years.

In Ralph of Laon's time, a good mathematician was still known as an *abaci doctor*, an "abacist," or even a "gerbercist." But monastery schools were slowly changing the way they taught arithmetic. Instead of shuffling counters on an abacus board, they were beginning to calculate with pen and parchment or a stylus on a wax tablet. They called this system the "algorithm," since the first complete Latin translation of al-Khwarizmi's *On Indian Calculation* was known by a Latinized form of its author's name, *Algorismus*. The introduction of the algorithm can also be linked to Gerbert and his students. Al-Khwarizmi's translator was the English monk Adelard of Bath, who was in Laon at the same time as Ralph. Adelard also wrote a treatise on the abacus, in which he explicitly credits Gerbert's influence on his work.

The step from Gerbert's abacus to Adelard's algorithm was very small. The main difference between the two ways of calculating was that the pre-drawn column lines of the abacus board disappeared; they were no longer needed once the place-value system was fully understood. In the algorithm, placement on the page alone distinguished a one from a ten or a hundred, and the use of zero to fill the empty space became standard. For a thousand years, we have added, subtracted, multiplied, and divided essentially the same way Gerbert taught his students at the cathedral school in Reims.

Outside the Church, the "nine signs" Gerbert introduced took much longer to catch on. In 1228, the famous mathematician Leonard of Pisa, known as Fibonacci, used al-Khwarizmi's *Algorismus*, along with other works translated out of Arabic, to write his *Book of the Abacus*. Together with Alexander de Villa Dei's *Song of the Algorithm* and John of Sacrobosco's *Popular Algorithm*, these thirteenth-century works made the use of Arabic numerals standard in teaching math. By 1270, the *Algorismus* had even been translated into Icelandic.

But the new system was not accepted everywhere. In 1299, the Guild of Money Changers in Florence issued an edict banning the new "letters of the abacus": "Be it stated that nobody . . . dare or allow that he or another write or let write in his account books or ledgers or in any part of it in which he writes debits and credits, anything that is written by means of or in the letters of the abacus, but let him write it openly and in full by way of letters." If this sounds extreme, remember that every time we write a check, we state the dollar amount both in Arabic numerals and then "in full by way of letters."

Fraud and mistakes were common with the new numbers. The versions found in Catalonia in the tenth century, and that Gerbert brought into the Christian West, are different from those used in Baghdad or in Constantinople: Gerbert's 5, 6, and 8 may have come from Visigothic script. But the numerals on the abacus board at Trier don't even exactly match the numerals that go with the mnemonic poem in the same manuscript. The 3 in the poem is upside down, the 4 has an extra loop, the 9 is lying on its back. Nor are all of the numbers recognizable to the modern eye: The 2 is almost always upside down (the numbers were written on a circular counter, so who could say which way was "up"?). The 3 looks like a backward Z with a hook. The 4 is more like a pair of swimming goggles and a snorkel than anything approaching a number. Worse, the number 5 looks very much like our symbol for 4. Until the printing press arrived in the fifteenth century, the shapes of the symbols were not standardized. It would take nearly five hundred years for Western Europe to realize the vision of a new arithmetic that Gerbert saw when he introduced his abacus board and its counters marked with nine Arabic numerals and zero to Reims before the year 1000.

CHAPTER VI

Math and the Mind of God

The Church found instant uses for Gerbert's abacus. Churchmen always needed to calculate the tithe—10 percent of a farm's proceeds—as well as other taxes, tolls, fines, and fees they could collect.

This "business math" had long been taught through story problems, such as the familiar goat-wolf-and-cabbage exercise: "A man needs to take a goat, a wolf, and a cabbage across a river, but his boat carries only two at a time. How can he get them all across unharmed?" Sometimes, the goat, the wolf, and the cabbage were strangely transformed. In one monk's notebook, the problem reads: "There were three men, each having an unmarried sister, who needed to cross a river. Each man was desirous of his friend's sister. Coming to the river, they found only a small boat in which two persons could cross at a time. How did they cross the river, so that none of the sisters were defiled by the men?"

In the late 700s, Charlemagne's schoolmaster, Alcuin of York, compiled a popular set of these puzzles. Its fifty-six problems are a portrait of monastic life: How many stones will it take to pave the cathedral floor? How many casks of wine will fit in the wine cellar? How many cowls can be sewn from a certain length of cloth? How many eggs will the monks eat for supper? In addition to problems of ordering, like the goat-wolf-and-cabbage one, they include problems with one or more unknowns: Your ox has been used to plow a field all day. How many hoof prints does it leave in the last furrow? You have a hundred silver coins with which to buy a mixed flock of a hundred chickens, ducks, and geese, each of which is priced differently. How many of each kind of bird will you buy? There are also arithmetic and geometric series: For example, a king raises an army by going to thirty villages and from each village brings out twice as many soldiers as went in. Solving such puzzles was something monks did for fun, like we do Sudoku. It was also a way to practice their times tables.

The basic textbook on arithmetic in Gerbert's day had been written in the 450s by Victorius of Aquitaine. Victorius had multiplied every number from 1,000 down to 1/144 by every number between 2 and 50. His massive times table was copied, recopied, studied, and memorized (at least partly) for over five hundred years. A popular handbook to the table was in progress at Fleury when Constantine wrote to Gerbert asking for the rules of the abacus. The *Commentary on Victorius's Calculus* by Abbo, then the schoolmaster at Fleury, was finished in about 982. Abbo discusses multiplying whole numbers and fractions, finger counting, and questions of weights and measures, such as why "a torch which is half-burned and thrown into water will surface with the burnt end uppermost," or why "the same quantity of honey will weigh half as much again as that of oil." Into the section on multiplication, Abbo slips mention of Gerbert's abacus and introduces the place-value system of calculating. That he makes no mention of Gerbert himself is not surprising.

Abbo and Gerbert were lifelong enemies. While Gerbert was schoolmaster at Reims, he sent Constantine a new set of mathematical exercises. He writes, "We are entrusting your sagacity, which has always flourished in the freest honesty of studies, with a prepublication of these axioms designed for the utmost exercise of the mind." The axioms, unfortunately, have been lost, but Gerbert says they pertain to digits and articles, and so to the abacus. "By using them," he continues, "the way for grasping these ideas is immediately opened to those persons of less comprehension who, because this pattern of thinking has either been neglected or completely unknown, exasperate every one of the skilled masters of the subject, moreover, by their habitual loquacity, replete with fallacies."

Who is this loquacious person whose understanding of the abacus is replete with fallacies? He is likely the same person to whom Gerbert was alluding when he sent Constantine his *Book on the Abacus*: that "half-educated philosopher" who "though really still a learner along with me . . . pretends that only he has knowledge of it." Abbo of Fleury calls himself *abaci doctor* in a silly little alliterative verse: *His abbas abaci doctor dat se Abbo quieti* ("With this, Abbot Abbo, abacus expert, rests"). But his writings do not reveal a very deep understanding of the new math.

Abbo was born within a few years of Gerbert in Orleans and was given to the monastery of Fleury as an infant. Like Gerbert, he was not noble, and like Gerbert he had an inquiring mind. Their curiosities—though not their personalities—overlapped. Abbo served as schoolmaster at Fleury while Gerbert was schoolmaster at Reims. Later, Abbo became abbot of Fleury just before Gerbert became archbishop of Reims—an appointment Abbo fought strenuously, as we will see, acting as the lawyer for Gerbert's rival, sending missives to the pope calling Gerbert a usurper, and trying to insinuate himself between Gerbert and his patrons, the king of France (successfully) and the Holy Roman Emperor (unsuccessfully). He made life miserable for Gerbert's beloved Constantine, who had replaced Abbo as schoolmaster at Fleury for two years while Abbo taught at Ramsey Abbey in England, and who had hoped to become abbot of Fleury himself. Despite Gerbert's string-pulling attempts, Abbo got the post. When he returned from England, Abbo had brought magnificent gifts: a gold chalice and vestments, gold bracelets, necklaces, and a large sum in silver coin. None of this hurt his chances at being elected abbot, though his medieval biographer says "some of the brothers perversely resisted the election."

Fleury under Abbot Abbo drew as many or more wandering scholars as Reims. English, Irish, and German monks visited there and left records of their stays, as well as donating books to the monastery's library and treasures to its altars. One medieval visitor claimed Fleury held three hundred monks. If that is true, it was the largest monastery in all of Europe. Why was it so popular? It housed the relics of Saint Benedict. The bones of the founder of the Benedictine order had been moved from Monte Cassino, where he had died, to Fleury in the late seventh century—or so the monks of Fleury claimed. The monks of Monte Cassino denied it, but they were shouted down.

Fleury was a center of the monastic reform movement, with Abbo a champion of monks' rights. He acknowledged only two authorities, the king and the pope. Bishops and archbishops, Abbo believed, should have no rights to the income of a monastery, and he fought hard to separate Fleury from the bishopric of Orleans. A dispute over a vineyard led to a pitched battle; swords were drawn, blood was spilled. Another time, a fight broke out at a church council: Abbo and the entire monastery of Fleury were excommunicated, and Abbo was called before the king to apologize for fomenting rebellion.

To achieve the same goal less violently, Abbo introduced a new way to keep track of the monastery's hundreds of charters and deeds for small pieces of land. Rather than storing the original parchments in boxes, bundles, or rolls, he had them recopied into ledgers—which made it, incidentally, easy to insert forged "documents" into a sequence. A forgery of a papal bull supposedly from the 750s, for example, convinced the pope in 997 to make Fleury fully independent from the bishop of Orleans. No bishop could say Mass, ordain a priest, or even enter the monastery grounds without the permission of the abbot—Abbo had won. Seven years later, he would lose his life in a brawl that broke out in a monastery whose monks did not care to be "reformed" by him. It's likely he was stabbed in the back.

According to his medieval biographer, Abbo studied at Reims before he became schoolmaster at Fleury in 975. This may be the key to his and Gerbert's mutual dislike, for Abbo was a failure as Gerbert's student. Abbo already considered himself an expert in grammar, dialectic, and arithmetic, and "was seeking to add other arts to his talents," wrote his biographer. "So he went to Paris and Reims, to those who taught philosophy. Under their guidance, he did indeed make a little progress in astronomy, albeit not as much as he had hoped." Returning to

Fleury, he paid a monk "much money" to teach him music. He remained unsure of himself in rhetoric and geometry (two of Gerbert's best subjects), but with five of the seven liberal arts under his belt, he still considered himself better than most of his contemporaries.

In the eyes of posterity, he does come out ahead of Gerbert: Though both were known as mathematicians, Abbo was proclaimed a saint; Gerbert was said to have sold his soul to the devil. And while Gerbert's genius has to be inferred, Abbo was prolific. Arguing that writing "did most to bridle the lusts of the flesh," Abbo wrote not only the book on calculation, but astronomical treatises, texts on logic and grammar and the calendar, a biography of the martyred King Edmund of East Anglia, a book of church law, an abridged *Lives of the Popes*, several poems in which names or double meanings are hidden in acrostics, and a collection of letters (each of them a political tract).

Some sense of what he taught as schoolmaster of Fleury can be gathered from the handbook written in 1011 by his student Byrtferth of Ramsey. There are passages on calculating; bits of natural history, grammar, and rhetoric; commentaries on the Venerable Bede and other authorities; and exhortations against the evils of the world. There is nothing new; it is all disappointingly derivative. Byrtferth's *Manual* summarizes the standard teaching textbooks of the previous centuries— and generally leaves out the more thought-provoking sections.

Abbo's own scientific works follow much the same pattern. They are clear, well-organized rearrangements of the sources commonly used in monastery schools over the hundred years or so that preceded his day. Rather than introducing a new scientific tradition, as Gerbert tried to do, Abbo was content to create a fine and tidy summation of the old one.

Both Abbo and Gerbert were known during their lifetimes as mathematicians. Why Abbo was made a saint, while Gerbert was called the devil's tool, has to do, not with their facility with numbers, but with how they presented their skill. Gerbert loved math for math's sake—as a way to stretch his mind. He saw God in numbers. Abbo put math to use in the service of the Church. He saw the abacus as a way to improve *computus*.

This kind of math had been at the center of ecclesiastical matters for hundreds of years. "Take number away, and everything lapses into ruin," complained Isidore of Seville in the 630s. "Remove computus from the world, and blind ignorance will envelop everything, nor can men who are ignorant of how to calculate be distinguished from other animals."

Computus comes from *computare*, "to count on one's fingers." In the Church, the term was even more specific. Computus was the technology needed to make a calendar—and the key to the Christian calendar was Easter. To begin the forty-day fast of Lent at the proper time, the date of Easter needed to be known well in advance. But Easter must fall on a Sunday, during certain days of the Jewish festival of Passover, and under the right phase of the moon, after the vernal equinox, but not during the raucous festival celebrating the founding of Rome.

Defining Easter meant integrating the lunar month (twenty-nine days, twelve hours) with the twelve months of the solar calendar, established by Julius Caesar before the birth of Christ. Caesar chose twelve to match the twelve signs of the zodiac through which the sun passes in a year. But 29½ times 12 is only 354, and the Julian calendar had 365¼ days—the quarter becoming February 29 every fourth year, or leap year. To get sun and moon in sync, you have to add an extra lunar month every three years.

Dating the vernal equinox was another problem—some churches used March 21, others March 25. And did a day begin at dawn, noon, sundown, or midnight? Then, on top of these natural cycles, governed by sun, moon, and stars, there was the week. The week was created by God. There is nothing natural about a seven-day cycle.

To make a calendar thus required a great deal of calculation. Monks like Abbo painstakingly calculated Easter tables for 19, 84, 95, 112, or even 532 years, each of which was considered by one school or another to be a full "lunar cycle," after which the date and the phases of the moon repeated exactly. Calculating the date of Easter, manuscripts show, was the most common math problem in a monastery school.

The most common error was losing track of the quarter days that add up to a leap day. Before Gerbert introduced Arabic numerals, all the monks' calculations were done in Roman numerals. Even after the abacus began to circulate, they continued to use Roman fractions, as Abbo's *Commentary on Victorius's Calculus* shows. Roman fractions are not decimal fractions, but base twelve—there are twelve parts, each with its own name, to a single unit. If you take an *uncia* (one-twelfth) from the whole, you are left with a *deunx* (eleven-twelfths); if you take a *sextans* (two-twelfths), you are left with a *dextans* (ten-twelfths). To make things worse, each uncia contained twenty-four *scripuli*, or, literally, "little stones." These clumsy fractions made it even easier to forget that, for every day, both sun and moon had to be accounted for. If you added a leap day to the solar cycle, you also had to add a day to the lunar cycle.

Easter tables were laboriously copied from manuscript to manuscript: 1,200 such copies, made from the eighth to eleventh centuries, still exist. Display models were carved in stone. The bishopric of Ravenna owned a sixth-century Easter table that Gerbert might have seen in later years. It was made of white Grecian marble, three feet square and over an inch thick, on which five cycles of nineteen years were laid out in the form of a nineteen-spoked wheel, radiating from a central cross. Yet in spite of such efforts, errors in calculation (or copying) often led to different dates for Easter at different churches in a given year.

While he was compiling an Easter table covering five future nineteen-year cycles, a monk named Dennis the Humble invented the concept of Anno Domini (A.D.)—the "Year of Our Lord"—the root of our modern dating system. In Dennis's time, years were dated by reference to the reign of the Emperor Diocletian. Using rules of thumb to calculate backward, Dennis decided Christ was born two hundred and forty-seven years before Diocletian took office, making the dates of his new Easter table A.D. 532 to 626.

Dennis's Anno Domini scheme did not immediately catch on. But two hundred years later, the Venerable Bede in England incorporated it into his book *On the Reckoning of Time*. From there it was picked up by Charlemagne, for whom it solved a pressing problem. According to the Church, the world would end six thousand years after the Creation. But 6000 Annus Mundi ("Year of the World") was the year Charlemagne was to be crowned emperor in Rome. He found it much preferable to change the date of his crowning to 800 Anno Domini. His change was codified in a new computus textbook by his court mathematician, Hraban Maur. (It begins beautifully: "Time is the motion of the restless world and the passage of decaying things.") Like Charlemagne, Hraban Maur left it to his successors to deal with the End of the World predicted to arrive a thousand years from the birth of Christ, in A.D. 1000.

As that date approached, Abbo of Fleury took up Hraban Maur's challenge—to move the problem of the Apocalypse into someone else's lap. He checked the calculations of Dennis the Humble and the Venerable Bede (possibly using a Gerbertian abacus) and found mistakes. If Saint Benedict had really died on an Easter Sunday—and assuming Abbo's Easter tables were correct—Bede's calculations were off by twenty years. And where was Christ's infancy? Dennis the Humble had followed Roman practice, counting Christ's birthday as the first day of the year A.D. 1. Abbo counted from Christ's birth as we count birthdays today: The birth is zero; on the first birthday, the child is one. According to Abbo's figures, then, the year 979 was actually a thousand years after the birth of Christ. Abbo spread the good news: A.D. 1000 had already passed without mishap; the predicted Apocalypse had not come. Abbo devised a new calendar, but no king or emperor or pope promoted his discovery, and our modern dates are still based on those of Dennis the Humble.

For all its practicality, math for Abbo was infused with spiritual purpose. Contemplating "what is unchangeable and true," he says in the introduction to his *Commentary on Victorius's Calculus*, "reforms the image and likeness of the Creator in man's soul." It provides "a defence against evil and error" and "leads men to God, who is himself Wisdom, by drawing them from the visible through the invisible to the unity of the Trinity." Arithmetic was thus a form of worship, leading one to recognize that "all number and mutability" derived from "unchangeable unity." Unity, he continued, "is a term of number from which 'one' is derived." And, rather less clearly, "What is 'one' 'is,' and what is, is one." One, for Abbo, was a symbol for God.

Abbo got these mystical ideas about the number one from Boethius, but they originated with the pagan thinker Pythagoras in the sixth century B.C. In addition to devising a theorem for the length of the long side of a right triangle, Pythagoras believed numbers had spiritual properties. Plato, in *The Republic*, picked up on the Pythagorean idea two centuries later, saying it was "good for the soul" to contemplate the "eternal verities as expressed by the properties of numbers."

Nicomachus of Gerasa, in the second century A.D., provided the next step. Nicomachus saw numbers in two ways: as both purely mystical (as in his book *Theologumena*) and more worldly (as in his *Introduction to Arithmetic*), but still not in any way approaching practical business math, what the Greeks called "logistics." Nicomachus's "arithmetic" is what we would call number theory: even and odd numbers, prime numbers, perfect numbers, numerical ratios or harmonies, polygonal and polyhedral numbers, and the three means (arithmetic, geometric, and harmonic). His book—well organized and clearly written—brought him fame. The name Nicomachus was to arithmetic as Euclid was to geometry.

In the sixth century A.D., Boethius translated Nicomachus's *Intro-duction*. He left "nothing essential" out, but, he says in his introduction to the work, "I did not restrict myself slavishly to traditions of others, but with a well formed rule of translation, having wandered a bit freely, I set upon a different path." He added quite a bit of geometry (Book Two is almost entirely taken up with triangles, squares, pentagons, hexagons, pyramids, cubes, spheres, and their proportions) as well as dwelling on the concept of unity, or how every "inequality proceeds from equality" and "every inequality can be reduced to equality."

Found in over two hundred medieval manuscripts, Boethius's *On Arithmetic* was taught in cathedral schools and universities throughout the Middle Ages and into the Renaissance. It had nothing to do with calculating the tithe or taxes, with Easter tables or anything practical. Why was it so popular? Because Boethius had Christianized Pythagoras. "Everything," Boethius writes, "which has been built up from the first substance of matter seems to be found in accord with the science of numbers. Therefore this was the original pattern in the mind of the creator." His biblical source is the Book of Wisdom, chapter 11, verse 21: "Thou hast ordered all things by number, measure, and weight."

This one line—illustrated by God leaning down from the clouds with his compass in hand (a visual tradition that led to William Blake's famous etching *Ancient of Days*)—justified the study of math and science in monastery and cathedral schools for hundreds of years. There was no clash between science and faith: Science *was* faith. Then Martin Luther took the Book of Wisdom out of the Bible in the sixteenth century, relegating it to an appendix. It was deleted altogether in Protestant Bibles of the nineteenth century—which may be one reason why many Americans today consider science and religion antithetical: No longer does math reveal the mind of God.

When it came to Boethius's *On Arithmetic*, and that mystical concept of "unity," Abbo's and Gerbert's curiosities overlapped. Yet where Abbo found a path to virtue, "a defence against evil and error," Gerbert found wonder and joy.

To him, *On Arithmetic* provided a treasure-trove of thought experiments. He owned a spectacular copy—written in gold and silver inks on purple parchment—which he would later present to Emperor Otto III. When the teenaged emperor asked him to explain it, he replied happily, "Unless you were not firmly convinced that the power of numbers contained both the origins of all things in itself and explained all from itself, you would not be hastening to a full and perfect knowledge of them with such zeal."

Gerbert delighted in the logical problems Boethius posed—and particularly in those that baffled more practical people like Abbo. For example: Take a ratio of three numbers, 16, 20, 25, and reduce it to the series 1, 1, 1. To Constantine, Gerbert writes, "This passage, which some persons think is insolvable, is solved thus. . . ." Following Boethius's rules for superparticulars (that is, "a number compared to another in such a way that it has in itself the entire smaller number and a fractional part of it"), through several pages of argument, Gerbert transforms 16, 20, 25 into 1, 4, 16, then 1, 3, 9, then 1, 2, 4. To resolve this so-called sesquiquarta into three equal terms, he explains, "take away the lesser from the middle, that is, 1 from 2, and place this 1 as the first term, and the remainder place second, that is 1. From the third term, that is from 4, take away unity, that is, 1, and twice the second term, that is two unities, and the remainder will be for you one unity." Gerbert concludes, "Therefore, you see how the whole quantity of the sesquiquarta has been changed into three equal terms, that is, unities: 1, 1, 1, not confusedly but in definite order, just as it was procreated in the beginning. This, therefore, is the true nature of numbers."

Baffling indeed. The problem seems meaningless. Yet to Gerbert, reducing 16, 20, 25 to three ones revealed the mathematical principle behind the creation of the universe. In the beginning all was One. From One, all of creation arose, logically and mathematically, and so—logically and mathematically—anything, no matter how complex, could be reduced once again to one. Figuring out how to do so was like reading God's mind. Gerbert's answer to this problem became so well known it was given a name, *Saltus Gerberti*, or "Gerbert's Leap," and sparked two of his contemporaries to try their own solutions: Notger of Liege and, of course, Abbo of Fleury.

This Boethian quest for unity can also be seen in a treatise on the physics of organ pipes, found in a twelfth-century manuscript alongside some of Gerbert's correspondence and Boethius's works on arithmetic and music. Called *Rogatus* from its opening words, *Rogatus a pluribus*, "having been asked by many," it is, like all of Gerbert's technical writings, a response to a student's question. It uses Arabic numerals (mixed with Roman numerals), well-honed Latin rhetoric, and quotations from Boethius, Pythagoras, Macrobius, Calcidius, Plato, and others. It is the work of an excellent mathematician, not only someone who understood the casting of metal pipes and the construction of organs. For these reasons, it has been attributed to Gerbert.

The *Rogatus* explains the physics of organ pipes in comparison to the more familiar monochord. The monochord is a single string stretched over a sounding box, rather like a one-stringed guitar. A movable bridge—like the guitarist's kapo—let the string be shortened to alter the pitch. When the string was "open" (no bridge) and plucked or bowed, it played a certain note. When the string was halved, using the bridge, the note was an octave higher.

The monochord was a favorite visual aid of Gerbert's when he taught the quadrivium. Music was part of the mathematics curriculum because all sequences or harmonies were translated into numerical ratios based on the monochord: The octave was 2:1. Music was a matter of numbers—even today we speak of fifths and fourths and diminished sevenths, time signatures, tempos, and rhythms. But to Gerbert and his peers, certain ratios of notes or rhythms were not just more pleasing to the ear, they were more sacred.

This theory was also based on Boethius. Sound, Boethius wrote in his equally mathematical *On Music*, was caused by percussion—the force of a vibrating string hitting the air. If the string was taut and beat the air rapidly, the sound would be high-pitched; if the string was loose and vibrated more slowly, the sound would be low. Sound traveled as a wave: Just as a pebble dropped into a pond causes rings of waves to spread out from the spot, so sound waves spread and grew fainter the farther they traveled from the vibrating string.

From this accurate scientific beginning, Boethius turned mystical. Sound—as music—was all around us. We heard the *musica instrumentalis*, music of the human voice and other instruments. But it was only a faint echo of the Music of the Spheres, the *musica mundana*, produced by the turning of the invisible spheres that held the stars (or perhaps by the spirit blowing through them). It was music, too, the *musica humana*, that held body and soul together. Because of this, we could be se-

duced to evil by immoral music, and restored to health, physical or spiritual, by music that was modest or simple (considered masculine), not
violent or fickle (feminine).

Composing modest, moral music was central to Gerbert's world. In
a monastery's seven services a day, much of each service was sung, or
rather, chanted, for the only sacred music of the time was the Gregorian
chant. Originally chant had no harmony: It was pure melody. With rich
and subtle variations of the melodic line, it required the hearer to listen
horizontally, across time, to recognize patterns and notice when they
shifted. When harmony began to be added, composers wondered why
some intervals made a "sweet mixture" and others sounded harsh. The
author of a musical tract from A.D. 860 concludes that we will never
understand the "deeper and divine reason" for this, since it "lies hidden in the remotest recesses of nature." Tellingly, he cites Boethius, "in
which it is convincingly shown . . . that the same numerical proportions by which different tones sound together in consonance also determine the way of life, the behavior of the human body, and the
harmony of the universe."

Gerbert had no interest in writing music. But he was fascinated by
the "deeper and divine reason" why some harmonies sounded sweet and
others harsh. According to Richer of Saint-Remy (who himself was a
cantor, or choirmaster), Gerbert was already a master of *musica* when he
left Spain. What Richer means by *musica* becomes clearer in his next anecdote: On his way to Reims for the first time, Gerbert instructed the
schoolmaster Gerann in mathematics. "But the difficulties of that science so discouraged him that he renounced completely the study of
music." Gerann did not, of course, renounce singing in choir—no
churchman could do that. But not everyone needed to know the un-
derlying mathematics of the chant.

Describing Gerbert's teaching methods, Richer writes, "Music, previously unfamiliar to the Gauls, he made very well known. Arranging
its notes on a monochord; dividing the consonants and symphonies
of the [notes] into tones and semitones, also di-tones and quarter-tones;

and dividing rationally the tones into sounds, he rendered [music] fully accessible."

At some point, it occurred to Gerbert to try to do the same with a pipe organ. The string of a monochord could be divided into halves, thirds, fourths, fifths, and sevenths. Plucked, it would produce the note of the scale the musician expected. But organ pipes, Gerbert noticed, were not like strings. A pipe half as long as another did not produce a tone a full octave higher. Pipes built using the Boethian ratios applicable to a string would produce acoustical distortions—the music would sound "off." Instead, the pipe maker had to add a little bit to the length of each pipe to tune it.

In his *Rogatus*, Gerbert explains mathematically how to compute the length of organ pipes for a span of two octaves. His solution was ingenious, though labor intensive, and stands up to the scrutiny of modern acoustics theory.

Yet Gerbert was not writing for organ builders who wanted their pipes to sound sweet. Instead he was searching for a mathematical truth: a law for computing the dimensions of an organ pipe that would sound the same note as the string of a certain length on the monochord. He came up with such an equation using what physicists call "opportune constants" (or "fudge factors") that allowed him to switch, mathematically, from the monochord to organ pipes and back. His treatise shows an extraordinarily modern perspective. He did not simply theorize—or search out authorities. He did experiments. He collected data and made practical acoustical corrections.

But behind his modern scientific approach remains a very medieval urge. Gerbert was searching for another form of "unity," like the 1–1–1 he had reached in the *Saltus Gerberti*, proof of the unity of Creation. His scientific goal, as always, was to reveal the single mathematical order in nature, given by God.

Gerbert also took a modern scientific approach to geometry. At the same time, he saw its utility in terms of religion. "It is full of

A page from Gerbert's geometry textbook, in a twelfth-century copy. The most advanced geometry book in the West at the time, it contains more Euclid than we would expect a tenth-century monk to know. This manuscript also contains a treatise on the astrolabe and a copy of one of Gerbert's mathematical letters.

accurate observations," he wrote, "for the purpose of . . . contemplating, admiring, and praising the wondrous meaning of nature and the wisdom of its Creator."

The textbook Gerbert wrote for his students at Reims was the most advanced geometry book in the West. It was not supplanted until centuries later, when Euclid was fully translated from Arabic (the Greek had long been lost) in the twelfth century. Gerbert's book contained, however, more Euclid than we would expect a tenth-century monk to know. One of his sources is known as *Geometry I*. It is attributed to Boethius, though the version that has survived is a clumsy concoction drawn from several works. Only one of them was Boethius's (now lost) Latin translation of Euclid. The copy Gerbert used still exists, in the library of Naples, with corrections made by Gerbert himself or one of his students. It contains the complete text and diagrams for numbers one through three of Euclid's definitions, postulates, axioms, and propositions; enough for a student to learn how to verify a theorem. There's also a good deal of the rest of Euclid's Book One, as well as lengthy excerpts from his other four books.

Gerbert drew from other sources as well. To Euclid, Gerbert added the geometrical bits of Boethius's *On Arithmetic* and his commentary on Aristotle, practical examples from the Roman surveyors' tradition (whose straight roads and arched aqueducts were used and admired in Gerbert's day), spiritual explanations by Saint Augustine, and introductory material from the standard quadrivium texts by Calcidius, Macrobius, and Martianus Capella. He compared and contrasted their approaches in a sophisticated and well-thought-out volume that proves Gerbert had a better grasp of geometry than anyone else in his time.

Gerbert begins by showing how any solid body was composed of parts. Taking a body—one that, he stresses, can actually be seen by the "eyes of the flesh"—he reduces it to its surfaces, the surfaces to lines, and the lines into points. The point, the most basic component of a solid, was equal to the number 1 in arithmetic: It represented the unity behind all creation. Geometrical unity, as in the 1–1–1 exercise in arithmetic, is perceived by the "eye of the soul." It, too, was a glimpse into the mind of God.

The triangle, likewise, was enticing because it was the basis for all other shapes. Gerbert writes, "The triangle exists for this reason as the origin and, as it were, the element in angled figures, in that every one of these figures is composed from it and resolved back into it again."

In his last letter before being elected pope, on the eve of the year 1000, Gerbert wrote to his friend Adalbold of Liege about this "mother of all figures." He was following up on an earlier letter, now lost, discussing how to find the area of an equilateral triangle. Gerbert's original example, with sides measuring 30 feet long, would have an area of 390, calculated using one standard method, he remarks, but 465 if calculated another way. "Thus, in a triangle of one size only," Gerbert says, "there are different areas, a thing which is impossible. However, lest you are puzzled longer, I shall reveal to you the cause of this diversity."

First Gerbert notes that area is calculated using square feet, not linear feet or cubic feet. Then, using a triangle of more manageable size, with sides measuring 7 units long, he explains the two formulas in use at the time: One was arithmetical, the other geometrical. The arithmetical method was the one taught by Boethius; the result it gave for the area of an equilateral triangle with sides 7 units long was 28. The geometrical answer gave 21, which was correct. Boethius, whose authority on most subjects Gerbert would believe, was wrong because, when "the triangle touches only a part, no matter how small," of a square foot, "the arithmetical rule computes them as a whole." Gerbert even drew a figure to make his point.

Adalbold and Gerbert must have often talked about math. We know Adalbold's bishop, Notger, was mathematically inspired by the *Saltus Gerberti*, and that Gerbert sent Adalbold his copy of *Geometry I* after he had used it to put together his textbook for Reims. Their mathematical correspondence would continue after Gerbert became pope and Adalbold the schoolmaster of the monastery of Lobbes (after Gerbert's death, he would become the bishop of Utrecht). In the only other letter that still exists, Adalbold asks the pope about finding the volume of a sphere.

The triangle letter is important not because the math is insightful, but because it exists at all. Churchmen—monks, clerics, bishops, even

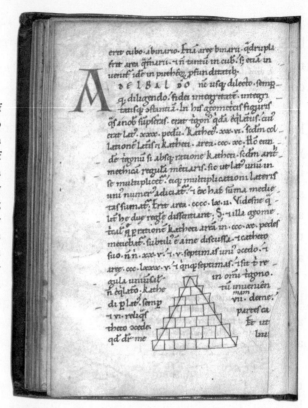

The beginning of Gerbert's letter to Adalbold of Liege on how to find the area of a triangle, from a twelfth-century manuscript containing Gerbert's geometry book. Mathematicians have found fault with the drawing, which may be different from Gerbert's original.

a pope—in the Dark Ages were investigating geometric puzzles, and they were doing so experimentally. Gerbert and Adalbold had noticed that their textbooks gave two different solutions to the same problem, which made no sense. So they experimented; Gerbert found the right answer—21—and explained the difference to his friend. The two schoolmasters did not simply accept the authority of their books, they questioned them and worked the answer out.

Gerbert found his solution by drawing a triangle with equal sides, each 7 units long. Then he cut out little squares of parchment, each 1 unit square, and laid them on top of the triangle. It took 28 squares to completely cover the triangle—as the arithmetical rule given by Boethius predicted. But parts of many squares stuck out over the lines. Copying this model into his letter, Gerbert explained to Adalbold, "The skill of the geometrical discipline, rejecting the small parts extending beyond the sides, and counting the halves about to be cut off and the [squares] remaining within the sides, computes what is shut in by the lines thus. . . ." After giving the formula (written out at length in words, not as we now write mathematics, as equations), he concluded, "To comprehend this more clearly, lend your eyes to it."

This concept of geometry as an experimental science caught on, at least in Liege. A famous series of eight letters exists from about twenty years later in which Rodolf of Liege and Ragimbold of Koln discuss the interior and exterior angles of a triangle; the definitions of linear feet, square feet, and "solid feet"; and how to find the correct ratio of the diagonal of a square to its side. They tried to understand theorems about the equality of angles or the sums of angles by cutting the angles out and laying them over each other.

A few years later, Franco of Liege tried to solve the famous puzzle known as "squaring the circle"—finding a square with the same area as an existing circle. He, too, began by cutting up a circle of parchment and trying to rearrange the pieces into a square.

The puzzle had been solved long ago by Archimedes. Franco knew that, but he did not know how: The Greek texts had never been translated into Latin. He had never heard of the idea of pi, without which he couldn't solve the problem. Nevertheless, he made a good effort. He was persistent, and set down his reasoning systematically, rather than just working at random. He also tried to solve related problems, such as finding a circle with the same area as an existing square. He came up with a powerful iteration procedure for finding square roots and showed that the square roots of 2, 3, and 5 could not be calculated as fractions, but could only be found geometrically. Though his work contains nothing new mathematically, it tells us that, in the early eleventh century, there was a vibrant school of geometry in Liege. This school existed as a direct result of Gerbert's teaching.

And yet, Franco *could* have known Archimedes' work. In 1999, a small thirteenth-century prayer book sold at auction for over $2 million. The manuscript was begrimed and moldy, some pages charred, some water-damaged, others stuck together. The prayers were almost illegible, the illuminations not very pretty (and later revealed to be forgeries painted after 1938). But prayers and pictures were not the point—the book was a palimpsest. The parchment had been reused—soaked in whey, the ink scraped off, the pages shuffled and turned ninety degrees to make a new half-size book. The erased text could still be partly discerned. It was a book of Archimedes in Greek.

The Archimedes codex contains the mathematician's well-known theorem on squaring the circle, explaining the concept of pi, along with his treatises on balancing planes, sphere and cylinder, measurement of the circle, spiral lines, and a game, known as *Stomachion* ("Bellyache"), in which fourteen cut-out shapes have to be reassembled into a square. It also contains works by Archimedes that modern mathematicians had never seen before 1999, such as Archimedes' letter to Eratosthenes, in which he explains his method. The *Method* combines calculus ("the mathematics of infinity") and physics. It was Archimedes' greatest achievement.

Based on the style of handwriting, scholars say that the Archimedes codex was written in Constantinople in about 975, possibly as late as 988—a time when numerous scientifically minded people in Gerbert's circle, from Reims, Barcelona, Rome, and even Cordoba, had contacts with the Byzantine Empire, and when books on science were being translated from Greek to Arabic to Latin. Gerbert was then the most influential mathematician in the West and one of the most avid book collectors. What would modern science look like, if the Archimedes codex had reached him?

CHAPTER VII

The Celestial Sphere

The school at Reims was famous for astronomy as well as math. This science Gerbert also pursued experimentally, creating instruments to observe, measure, and model the brilliant chaos of the starry sky.

That sky was not the few twinkles we're used to. Thanks to smog and light pollution, most of us have never seen the full shimmering panoply of stars, planets, and Milky Way. Modern astronomers associated with the Dark Sky project like to tell of the blackout following the Northridge earthquake in 1994. Los Angelenos lit up the hotlines, fearful of the "giant silvery cloud" over the city. To Gerbert, that cloud was a clock and a compass; its regularities (and irregularities) gave a lesson in divine harmony, a way to reach God by studying his Creation.

At its most practical, studying the night sky—what medieval astronomers called the Celestial Sphere—was a way to tell time. The Rule of Saint Benedict prescribed prayers at the first hour of the morning; the third hour of the morning; the sixth hour, or midday; and the ninth hour—all times that, in Rome, had been announced by the changing of the guard. Other prayers were said at sunrise and sunset, dawn and dark. These last four were not difficult to identify, but what was "the third hour of the morning" to a monastery in France, where guards were not changed with loud clashing of sword on shield and—this was the greater difficulty—the normal concept of an hour of daytime and an hour of nighttime were not even? An hour of daytime was not sixty minutes, but one-twelfth of the time the sun was up; an hour of night-time was one-twelfth of the darkness. A night-hour in the summer was thus significantly shorter than a night-hour in the winter.

In the late 500s, Gregory of Tours came up with a new way to know when to pray. The ordinary, unequal hours he called the *temporal* hours. But astronomers knew another measure of time: They divided the circular motion of the stars into twenty-four equal hours, which they called *equinoctial* hours, because only at the two equinoxes are days and nights (and thus hours) the same length. Gregory came up with a formula for calculating the average day-length—how long the sun shone in equinoctial hours—for each month. He converted that sum into temporal hours (dividing by twelve), rounded up, and adapted his sundial each month accordingly.

But what happened at night, when a sundial was useless? Gregory calculated the length of each month's average temporal night-hour, then counted how many psalms went into that hour. The monk on night watch would dutifully chant the required number of psalms, then ring a bell to wake his brethren—the original alarm clock (the word "clock" comes from *glocke*, German for "bell").

Gerbert taught a much more precise way to keep time. With the astronomical instruments he made, Gerbert could calculate, during the day, the time that a certain star would rise or set. Conversely, seeing a star rise (or set) at night, he could tell the time—in either temporal hours or equinoctial hours—to within a quarter of an hour. He could trace the path of any star and say where in the sky it would be at any time of night or day—and as seen from any place on earth. He could calculate the length of daylight (in equal hours) or the length of a daytime hour (in unequal hours) for any day in any place. He could find the height of the sun at noon for any day in any place, or, given the height of the sun at a certain place, and knowing the day of the year, he could tell the time.

Gerbert's instruments are known as celestial globes or celestial spheres, because they are models of God's Celestial Sphere. Richer of Saint-Remy, in his history of France, sketches four of them. After so tangling himself up in technical details that it isn't clear what any of them actually looked like, Richer—no astronomer, he—figuratively throws up his hands: "It would take too long to tell here how he proceeded further."

In his letters, Gerbert provides very little more. He talks mainly about the difficulty of making these spheres. Remi, the schoolmaster at the important cathedral of Trier, for example, had written to ask about a fine point of mathematics on the abacus, which Gerbert succinctly explained. Remi also wanted a celestial sphere. Gerbert replied:

> We have sent no sphere to you, neither have we any at present; nor is it an object of small work. . . . If, therefore, you are eager to have this that involves so much work, send to us a carefully written volume of Statius's *Achilleidos* in order that, unable to have the sphere gratis because of my excuse of its difficult construction, you may be able to wrest it from us as your reward.

Four months later, having received the first half of the poem about Achilles, he wrote again to Remi:

> Your good will, beloved brother, was overburdened by the work of the *Achilleidos* which, indeed, you began well, but left incomplete because your copy was incomplete. Since we are not unmindful of your kindness we have begun to make the sphere—a most difficult piece of work—which is now both being polished in the lathe and skillfully covered with horsehide. So, if you are weary from the excessive anxiety of anticipation, you may expect it, divided by plain red color, about March 1st. If, however, you are willing to wait for it to be equipped with a horizon and to be marked with many beautiful colors, do not shudder over the fact that it will require a year's work. As for giving and receiving among our followers, how true the saying that he who owes nothing need return nothing.

Gerbert was irked. Even Constantine had never asked for a celestial sphere, only the instructions for making one. But Remi could not be turned down. The cathedrals of Reims and Trier were closely linked—among Gerbert's letters are nineteen he wrote, under his own name or for Archbishop Adalbero, to Archbishop Egbert of Trier. One concerns the monk Gausbert, whose notes on Gerbert's abacus may now rest in the Trier archives. Other letters show the warm friendship between Adalbero and Egbert, both noblemen from Lorraine, both politically astute, both engaged in a massive program to enlarge and embellish their churches and to claim "primacy"—the right to sit down in the presence of the king or emperor before the other bishops.

Set amid the vine-covered hills of the Mosul wine region, Trier had been a Roman capital. Roman gates opened onto the medieval city, which was the oldest archbishopric north of the Alps. Egbert made much of the fact that Saint Peter himself had sent the first bishop there. Egbert owned a relic—a chip of wood—from the staff Saint Peter gave that bishop. He encased it in a six-foot-tall crozier coated with gold foil and encrusted with jewels, coins, enamels, and ivory work showing a distinct Byzantine influence. When the region suffered a drought, Egbert proceeded from church to church, brandishing Saint Peter's staff. The metalwork of Trier was so fine that Adalbero commissioned a jeweled cross for the Reims cathedral; he arranged to pick it up while visiting a holy place along the Rhine.

Gerbert accompanied Adalbero on that pilgrimage, during which they were forced to detour. Gerbert wrote to Egbert, saying, "Continual torrents are in possession of the slopes of the mountains. Ever-flowing waters so clothe the fields that, with villages and their inhabitants submerged and the herds destroyed, they bring terror of a renewal of the Flood. The hope for better weather has been shattered by the *phisicis*. Accordingly, we are fleeing to you just as Noah to the Ark."

The word "phisici" could refer to experts in physics, such as meteorologists—or to astrologers. Perhaps while they were waiting at Trier for the floodwaters to recede, the two schoolmasters, Gerbert and Remi, talked about astrology. In those times, the term encompassed both the scientific study of the stars and fortune-telling. A celestial sphere was useful for both.

A fifteenth-century wood carving from Ulm Cathedral showing what one of Gerbert's simplest celestial spheres might have looked like. The astronomer pictured is Ptolemy.

Remi may once have been Gerbert's student. That would explain why Gerbert was miffed at being asked to do the man's work for him. But Remi was also well known as a scholar: He wrote a book on the abacus, following Gerbert's, along with hymns and sermons and a *Life of Saints Eucharius, Valerius, and Maternus,* the founders of Trier, written in rhymed prose. He was named abbot of Mettlach before Egbert's death in 993—which, curiously, is just when the English monk Leofsin left that monastery, taking a copy of Gerbert's abacus with him to Echternach.

The celestial sphere Remi wanted was not Gerbert's invention. Such instruments were known since ancient times. Cicero mentioned them in his *Republic,* a book Gerbert asked Constantine to bring with him once when he visited Reims. Gerbert may also have read about them in Plato's *Timaeus,* through the commentary by Calcidius, a third-century author well known to him. In Martianus Capella's fifth-century handbook of the liberal arts, of which Gerbert owned a copy,

Lady Geometry bears a globe on which "the intricate patterns of the Celestial Sphere, its circles, zones, and flashing constellations, were skillfully set in place."

While celestial globes or spheres are referred to in books that Gerbert knew, instructions on how to make them are rare. Ptolemy gives explicit instructions on making these devices in his *Almagest*, a work that circulated in tenth-century Spain—but only in Arabic, as far as we know. A seventh-century Byzantine writer, Leontius, made such spheres out of wood, smoothed over with plaster and painted a dark blue. The Syrian astronomer al-Battani, whose father was an instrument maker, made an elaborate one of metal that he called "the egg"; his description was translated into Latin in Catalonia before the year 1000. In 1043 an instrument maker wrote of seeing a silver celestial sphere in the library of Cairo. It had been made by the Persian astronomer al-Sufi, who died in 986, and weighed as much as 3,000 silver coins. The oldest extant celestial sphere, dated to about 1080, is uniform and precise, evidence of a long history of sphere-making preceding it. Like most remaining spheres, it is made of metal, in this case brass, formed as two joined hemispheres.

Gerbert may have learned to make his celestial spheres of wood and horsehide in Catalonia, from Arabic sources (written or oral). Or he may have concocted them out of classical learning, hearsay, and his own ingenuity. From the description by Richer of Saint-Remy it is clear that nothing like them had ever been seen at Reims.

No one else writing in Gerbert's lifetime describes a celestial sphere. But two later treatises give practical tips on making them. First you need wood that will not warp, split, or rot. "It must be gathered when the moon is waning in the last days of the lunar month," says an Arabic text translated in Spain in the thirteenth century. Soak it in hot water for two days and dry it in the sun; if it warps or splits, start over. Turn the block of wood on a lathe until it is perfectly round. Carefully cut off a small circle and hollow it out, gluing the circle back as a plug. Cover it with leather "of the sort used for shield covers, but cut thinner." Then coat it with plaster and paint it as dark as night. The stars, says a Latin text printed in 1518, are made of wire pushed into the wood, flush with the surface.

How the constellations are laid out depends—literally—on your point of view. Since antiquity, there had been two ways of thinking about the stars. In each case, the stars were imagined as attached to a hollow sphere rotating around the earth at its center. You could stand on the earth and look up at the stars. (This is the viewpoint of a modern planetarium.) Or you could imagine yourself hovering above the sphere of the heavens, like God, looking down through the stars to the earth. Illustrations in medieval astronomy books show both points of view: If you looked up, the figures of Orion or the Great Bear faced you; if you imagined yourself looking down, they would turn their backs to you and become reversed, right to left. All extant celestial spheres take the God's-eye view.

To fix the stars in their proper places, according to al-Battani, the sphere needed three great circles: the equator; a circle perpendicular to the equator and passing through both poles; and the ecliptic, the path the sun seems to take through the sky, inclined at an angle of 23½ degrees from the equator. To draw the circles, Gerbert used a pair of compasses. He divided each circle into 360 degrees. To set each star, he observed (or calculated) its celestial coordinates: its distance in degrees from the equator, the north-south line, and the ecliptic.

Once the stars were in place, other circles could be painted on. Parallel to the equator were the so-called climate circles: the Arctic Circle, the Tropic of Cancer, the Tropic of Capricorn, and the Antarctic Circle. The Arctic Circle marked the latitude where, for observers in the Northern Hemisphere, such as Gerbert, the same stars were visible all night; the stars that circled the Antarctic were always invisible. The tropics—from the Greek *tropos*, or "turning"—are the circles made by the zodiac signs Cancer (the northernmost sign on the ecliptic) and Capricorn (the southernmost sign) as they turn around the earth each day.

The zodiac is a narrow band to either side of the ecliptic. It was divided into twelve signs, each spanning 30 degrees of space and marked by a constellation. The degrees were counted off from the vernal equinox, at the beginning of Aries, where the zodiac crossed the equator. The width of the zodiac band is defined by the travels of the moon. When the moon crosses the ecliptic—as the name implies—an eclipse can occur. The five "wanderers"—Mercury, Venus, Mars, Jupiter, and Saturn (the other planets were not discovered until after the telescope was invented in the 1600s)—are also confined by the band of the zodiac. Some medieval astronomers preferred to call the planets "confusers," from the Greek *planontes*, instead of "wanderers," from *planetai*. Like all of God's creation, planetary motions are by definition orderly. "There is no inconstancy in divine acts," said one early astronomer. If the planets appear to wander, our eyes have simply failed to perceive God's pattern.

Another circle was necessary to make the celestial sphere useful for telling time, but it couldn't be painted on: the horizon. On earth, the horizon is the line between ground and sky. In the heavens, the horizon divides the stars you can see from those that have not yet risen. To mimic the rising and setting of the stars, Gerbert let his wooden sphere rotate within a ring—probably brass—that represented the horizon.

To master the celestial clock, a monk first needed to know his constellations. To teach them, Gerbert made a second, simpler sphere, Richer says, with no circles, on which the constellations were clearly mapped in iron and copper wire. A sighting tube through the poles acted as the axis. By sighting on the North Star, the sphere could easily be aligned with the night sky. This sphere, says Richer, "has something divine in itself, as even those ignorant in this science, if they were shown one of the constellations, could then recognize all the other constellations thanks to the sphere and without the aid of a master."

The concept of the climate circles also needed to be taught. For this, Gerbert made another instrument, using a hollow sphere cut in half along the north-south axis. He added seven sighting tubes, one for the north pole, one for each of the five climate circles, and one for the south pole. Each tube was half a foot long. "They differ from organ pipes by being all equal in size, in order not to distort the vision of anyone observing the circles of the heavens," Gerbert explained in a letter to Constantine. To keep them steady, they were attached to an iron semicircle. To use the device, Gerbert wrote, aim through the two pole tubes at the North Star. Turn the hemisphere round side up and fix it in place. "You will be able to determine the North Pole through the upper and lower first tube, the Arctic Circle through the second, the summer circle [or Tropic of Cancer] through the third, the equinoctial circle [or equator] through the fourth, the winter circle [or Tropic of Capricorn] through the fifth, the Antarctic Circle through the sixth. As for the south polestar, because it is under the land, no sky but earth appears to anyone trying to view it through both tubes."

This instrument, says Richer, "was so well contrived that . . . it brought to light circles which were new to the eyes and securely fixed them deep in the memory." In addition to teaching the concept of the climate circles, it also allowed Gerbert to measure the height of a star above the horizon, at least approximately, in relation to those circles.

Gerbert's fourth and most sophisticated astronomical instrument was an armillary sphere made of seven open rings (in Latin, *armilla*). Two rings passed through the north and south poles, spread at a 90-degree angle to create the basic sphere shape. Perpendicular to these, Gerbert placed five rings, graduated in size, for each of the five climate circles. To the outside of this sphere of seven rings, Gerbert attached the oblique band of the zodiac. Inside the sphere, writes Richer, "he suspended the circles of the planets by a very ingenious mechanism. It could show, in a surprising manner, their absides, their altitudes, and their respective distances."

These three measurements revealed how the universe worked. In Gerbert's day, the sun, moon, and planets were thought to circle the earth in eccentric, not perfectly round, orbits. Earth was not at the center of most of these orbits. A planet could thus seem to speed up when it came closer to Earth and slow down as it drew farther way. The planet's orbit itself was called an "apsis" (plural "apsides"). A planet's "altitude" was its apogee, or farthest point from the earth. Its "distance" defined the interval between its orbit and its neighbor's.

Drawings from the ninth through eleventh centuries illustrate this theory—and its problems—clearly (see Plate 6). The circle of the moon is generally centered on the earth, while some of the other circles are more eccentric than others. One image shows the sun dead-center in the zodiac, with Earth well off to the side (though the lines show the sun still managing to circle the earth). Mercury and Venus also inspired serious debate in Gerbert's time: Several illustrations show these two inner planets circling the sun, which itself circled Earth. In a manuscript made at Fleury around the year 1000, the astronomer, clearly intrigued, has drawn three different diagrams to try to perceive how such epicycles could work: Are they two concentric circles? Intersecting circles? Arcs?

Gerbert explained the planets' wanderings to his students through three-dimensional models, not two-dimensional drawings. First, he made observations. Then he built an armillary sphere, like the one Richer described, to model the planets' movements. That his students picked up on his experimental technique is proved by a drawing discovered in 2007. As a theoretical model it is almost unreadable. To the uninitiated, it looks more like a ball of twine than a model of the heavens. The orbits of the planets are drawn, not as circles, but as wide bands overlapping each other in confusing ways—until you realize the picture is not a diagram but a drawing of an actual three-dimensional object. It is a picture of one of Gerbert's armillary spheres.

It does not exactly match Richer's description. The seven armillary rings themselves are missing. But these rings are structural; they do not affect the absides, altitudes, and distances of the planets. And even if he had wanted to add them, the artist was not technically skilled enough to do so: The concept of perspective drawing, which allowed artists to give the illusion of depth to flat representations of three-dimensional objects, was not invented until the 1400s.

The manuscript containing this illustration is a copy of a student's notebook—a mishmash of notes and drawings. The copy was made in Augsburg, Germany, in the twelfth century, and now resides in the Vatican archives. The teacher, unnamed, had devised a far-ranging course of study that matches exactly what we know about Gerbert. He used visual aids to explain mathematics (the abacus), astronomy (celestial spheres), and music (the monochord). He quoted classical authors such as Boethius and Pythagoras. Some of the same terminology and topics occur in the letters Gerbert sent to his students. Finally, the manuscript contains two different diagrams explaining the various subjects that make up philosophy. In one diagram, physics is shown as a subdiscipline of mathematics. In the other, the two are separate and equal sciences. Two diagrams like these, as we will see, sparked a famous debate in 980 between Gerbert and another schoolmaster who was jealous of Gerbert's fame. The original of this notebook could have belonged to the spy Gerbert's rival had sent to infiltrate the Reims school.

W ith his armillary sphere, Gerbert could demonstrate the relative movements of the sun, moon, and "wandering stars." He could explain eclipses, rotating the small wooden balls that represented sun or moon around on their metal rings until they lined up with the earth and then, with a candle, showing how blocking the sun's light caused the shadow.

Kings always needed someone to explain eclipses. Medieval historians unanimously called them "terrible" and "terrifying." The eleventh-century French historian Ralph the Bald described "a terrible event, an eclipse or obscuring of the sun from the sixth to the eighth hour. Now the sun itself took on the colour of sapphire, and in its upper part it looked like the moon in its last quarter. Each saw his neighbour looking pale as though unto death, everything seemed to be bathed in a saffron vapour. Then extreme fear and terror ripped the hearts of men."

The bishop of Liege was with the emperor on campaign in southern Italy in 968 when there was an eclipse. Called on to calm the soldiers' fears, says the *Life of Heraclius*, the bishop quoted "Pliny, Macrobius, Calcidius, and many others, both *astrologi* and computists," to explain the phenomenon scientifically.

Ralph the Bald must have been exposed to such a lecture. Later in his history he writes about "an eclipse of the moon which terrified men.

At the eighth hour of the night either God placed some wondrous thing between the sun and moon, or the sphere of some other star intruded into that position. What really happened is known only to the Creator. At first the whole moon took on a foul and bloody aspect, which went on gradually disappearing until dawn next day." Still later, after having spoken with an archbishop of Reims, Ralph has learned a little more: "Eclipse means a failing or a want, but it is not the result of any failing within the heavenly body itself, but occurs because it is hidden from us by some obstacle."

As long ago as 612, when Isidore, the bishop of Seville, sent his first draft of *On the Nature of Things* to his patron, the king of the Visigoths replied with a diagram explaining a lunar eclipse. He labeled the obstacle that keeps the sun's rays from reaching the moon as the *globus* of the earth. In fact, every explanation of an eclipse Ralph could have heard around the year 1000 required him to know already that the earth was round. No one conceived of the cosmos as a huge, vaulted arch with the earth below, flat as a plate and surrounded by ocean.

Gerbert's teaching that the earth was a globe was not heresy—as later interpreters would have it—but orthodox Catholicism. Saint Augustine himself, the most influential of the Church Fathers, noted in A.D. 401 that it was both "disgraceful and dangerous" if a Christian was heard "talking nonsense on these topics." How could pagans believe us on "matters concerning the resurrection of the dead, the hope of eternal life, and the kingdom of heaven," Saint Augustine said, if they think our holy books "are full of falsehoods on facts which they themselves have learnt from experience and the light of reason? Reckless and incompetent expounders of Holy Scripture bring untold trouble and sorrow on their wiser brethren," he concluded, "when they are caught in one of their mischievous false opinions."

We do not know which reckless and incompetent fools Augustine was thinking of. The argument over the shape of the earth has a long history. A thousand years before Augustine, Thales of Miletus had suggested the earth was a flat plate floating on the ocean. Thales's contemporary, Anaximander of Miletus, argued instead that it was a cylinder hanging in empty space. Pythagoras, born a hundred years later, around 530 B.C., decreed the earth was a sphere, because a sphere was a perfect shape. It and the other heavenly bodies—including the sun—circled around a central fire, he thought. Plato proposed that the earth was at the center of the universe, but later decided Pythagoras's central fire made more sense. Aristotle moved the earth back to the center and listed scientific observations to prove it was a globe, one being that the shadow of the earth on the moon during an eclipse was circular. By 240 B.C., Eratosthenes had devised how to calculate the spherical earth's circumference.

Few thinkers challenged Aristotle's model of the universe over the next thousand years (though Pliny, in the first century A.D., thought perhaps the earth was an irregular sphere, more like a pinecone). Most churchmen found the shape of the earth irrelevant; in the words of Saint Basil of Caesarea (A.D. 330 to 379), it is "of no interest to us whether the earth is a sphere or a cylinder or a disk, or concave in the middle like a fan."

But the idea of a flat earth had not entirely disappeared. One writer who argued in favor of it was Lactantius, born in Africa in about A.D. 245. He was a professional rhetorician, a witty writer, and a convert to Christianity who possessed the convert's virulent opposition to anything that smacked of paganism. If the earth were a sphere, he scoffed, the people on the bottom side would have their feet above their heads, trees would grow upside down, and rain would fall up from the sky. For this and other theories—he claimed, for example, that Christ and Satan were twin angels, one good, one evil, created by God to balance each other out—Lactantius was condemned by the Church as a heretic. Saint Augustine may have been thinking of him and his followers.

Another known Flat Earther was a Greek merchant and later monk, Cosmas Indicopleustes (the name means "Sailor to India"). He described heaven in A.D. 547 as a rectangular arch above a rectangular earth, flat as a plate and surrounded by ocean. At the northern end of the earth, he wrote, rose a mountain, which hid the sun at night. In his lifetime, Cosmas was dismissed as an idiot—his critics quoted Saint Augustine's complaint, from more than a hundred years earlier, against

Christians talking nonsense. Cosmas, however, was unknown to Gerbert. Cosmas wrote in Greek, and his cosmology was never translated into Latin during medieval times. It first came out in Latin in 1706.

In Gerbert's time there were no Flat Earthers to argue with. Gerbert and his peers followed the teachings of the Church, codified by Isidore of Seville and the Venerable Bede, who both described the earth as round like an egg. When Lady Geometry appeared in ninth- and tenth-century manuscripts of the popular schoolbook by Martianus Capella, the commentators writing the glosses between the lines made sure their students understood the shape of the earth. "I am called Geometry because I have often traversed and measured out the earth, and I could offer calculations and proofs for its shape, size, position, regions, and dimensions," says Lady Geometry, at which point one commentator interrupts: "It has a round shape." After explaining how Eratosthenes calculated the circumference of the earth, Lady Geometry continues to describe it: "Nor is it concave," she says. Adds this commentator, it is not "like a sponge, curved, but spherical, that is, like an egg." In the margin he drew a circle, labeled: "A disc. The earth is not like that."

The historian Ralph the Bald had never studied geometry. His education was limited, his Latin was poor. His point of view was overwhelmingly mystical: He attributed every act and event to the will of God. But even Ralph knew the earth was round. Describing the imperial insignia, he said it was "made in the form of a golden apple set around in a square with all the most precious jewels and surmounted by a golden cross. So it was like this bulky earth, which is reputed to be shaped like a globe."

Maps of the world before the year 1000 also depicted the earth as a sphere—if you know how to read them. Depicting a three-dimensional globe on a flat surface has always been problematic. Modern mapping conventions, for example, stretch the world into an oblong and cut off the far north and south. The tenth-century standard—in geometry texts as well as on maps—was to represent a sphere as a circle. With no concept of perspective drawing, the artist had little choice.

The earliest existing world map, from a chapter that Isidore of Seville added in 613 to *On the Nature of Things*, is a circle. It shows three continents—Asia, Europe, and Africa—encircled by ocean. In some copies of the map, Asia is at the top, Europe to the left, and Africa to the right. In others, with Asia still at the top, Europe is on the right, Africa on the left. Scholars have used this discrepancy to dismiss the entire map as nonsense. Yet what is usually pegged as an error proves instead that the mapmaker knew his astronomy: Just like putting the stars onto a celestial sphere, mapping the earth can be done from two perspectives.

One is the God's-eye view, made popular by a medieval textbook, the *Commentary on the Dream of Scipio* by the fifth-century writer Macrobius. Scipio is a character in Cicero's *Republic*. Transported to the stars in a dream, he looks down on the little ball of the earth below, its great cities reduced to tiny spots surrounded by desert and wasteland, the known world itself but an island in the vast ocean. The moral? Man's struggle for fame is meaningless. This perspective is the one modern maps are made from—and why we recognize the continents as seen from outer space. But there is another way. Feet firmly on the earth, looking up at the stars, the mapmaker can stand with his back to Asia and find Europe on his right hand and Africa on his left.

A world map that Gerbert might have seen was made in the monastery of Fleury while Abbo was abbot (see Plate 7). The manuscript that contains it collects several texts relating to computus, including some by Abbo himself. The map is drawn as a circle, from the God's-eye view, and accompanies selections from Macrobius's *Commentary*. A caption at the top says the circle depicts one hemisphere of the globe. Around the edge of the circle, another caption refers to the calculation of the earth's circumference by Eratosthenes.

Rather than being centered in the circle, as on Isidore's map, the known world here is scrunched into the top third, with Europe on the top left, Africa beneath it, and Asia on the right (very much like we would draw them). The Mediterranean and its islands divide the three. A band of ocean runs around the outer edge of the circle; little dots in the northern seas are labeled "Britannia," "Hibernia" (Ireland), and

"Tile" (Thule, or Iceland). Two bands of ocean cross the hemisphere, leaving a large continent straddling the equator; this "torrid zone" is said to be uninhabitable. But below the southern band of ocean is another crescent of land, unnamed. This was the land of the Antipodes. The caption says it is inhabitable—though perhaps no people are yet there. For if people lived in this unreachable land, separated from civilization not only by the ocean but by the uncrossable deserts surrounding the equator, how could they be descended from Adam and Eve?

In spite of this biblical difficulty, many writers in Gerbert's time believed there were strangers on the opposite side of the globe—and thought the Church should send explorers out to find them. Those who disagreed quoted Saint Augustine's well-known dismissal of the Antipodes theory. As he wrote in *The City of God:*

> It is not affirmed that this has been learned by historical knowledge, but by scientific conjecture, on the ground that the earth is suspended within the concavity of the sky, and that it has as much room on the one side of it as on the other: hence they say that the part which is beneath must also be inhabited. But they do not remark that, though it be supposed or scientifically demonstrated that the world is of a round and spherical form, yet it does not follow that the other side of the earth is bare of water; nor even, though it be bare, does it immediately follow that it is peopled.

And indeed, science could not say, in Gerbert's day, whether or not the unknown face of the globe was underwater or unpeopled. Those questions would not be answered, definitively, until Columbus discovered America five hundred years later.

Christopher Columbus is often given the credit for proving the earth was round, as well. Accepting that notion—as many modern textbooks do—means ignoring everything Gerbert knew and taught about the earth and the heavens. Such ignorance is not a product of chance: The Flat Earth Error, as it is called, was created.

The Error begins with the Italian poet Petrarch, who is known for two things: developing the sonnet, and coining the term "the Dark Ages." Sometimes called the first humanist, Petrarch divided history into ancient (before Rome became Christian in the fourth century) and modern (his own time, the fourteenth century). Everything in between was dark.

In Petrarch's version of history, the world suffered through a thousand years of ignorance and superstition. Then the humanists heroically resurrected the classical truths of Greece and Rome—their art, literature, philosophy, science—or so they wanted people to believe. They also had a political motive. The Italian cities wanted to break free of the Holy Roman Empire. That meant denying all the contributions to civilization that Gerbert's emperors (Ottos I, II, and III) had sponsored, as well as those promoted by Charlemagne, not to mention by the Church itself. Petrarch and his fellow humanists saw no contradiction in the fact that all of the ancient art and learning they "discovered" had been copied, and so preserved, in the scriptoria of monasteries and cathedrals through the thousand years of the so-called darkness. Instead of promoting Gerbert and his celestial spheres, they revived Lactantius and his rain that falls "up" from the sky.

By the 1700s, the Dark Ages were more politely known as the Middle Ages (Latin *Medii aevi*, from which we get "medieval"). In Protestant circles they still represented a blank spot of barbarism and superstition (i.e., Catholicism) between antiquity and the Renaissance. Henry St. John Bolingbroke, whose political writings influenced Thomas Jefferson, among others, called studying the Middle Ages "a ridiculous affectation in any man who means to be useful to the present age." This intellectual attitude made it easy for Washington Irving, in *The Life and Voyages of Christopher Columbus*, to write a revisionist version of the discovery of the New World in 1492.

In the 1820s, having just published the stories "Rip Van Winkle" and "The Legend of Sleepy Hollow" to popular acclaim, Irving went to Spain, where he was given access to original documents about Columbus. Finding the truth a little dry, he decided to embroider a bit on the historical Council of Salamanca, which had been convened to judge whether Columbus's proposed voyage to discover a western route to India was a good risk of the king's money. In the monastery of Saint-Stephen, "the most scientific college in the university," our hero "soon discovered that ignorance and illiberality may sometimes lurk under the very robes of science," Irving wrote. Facing a "learned junto" of professors, monks, and church dignitaries, this "plain and simple navigator, somewhat daunted, perhaps, by the greatness of his task" met their "mass of inert bigotry" with an "elevated demeanor" and a "kindling eye."

They threw the heretic Lactantius at him, Irving claims, as well as Saint Augustine's views on the Antipodes. "To his simplest proposition, the spherical form of the earth, were opposed figurative texts of Scripture," Irving writes: The Psalms and Saint Paul both describe the heavens as being like a tent, ipso facto the earth was flat like the floor of a tent. "Others," Irving concedes, "admitted the globular form of the earth, and the possibility of an opposite and inhabitable hemisphere, but maintained that it would be impossible to arrive there."

What in fact they maintained—and the original records of the council still exist—was that Columbus was fudging his numbers. Using the same methods Gerbert knew, the Council of Salamanca calculated the circumference of the earth to be about 20,000 miles (it is actually about 24,900 miles) and the distance between one degree of latitude or longitude at the equator to be 56⅔ miles (it is actually 68 miles). Columbus thought the earth was much smaller. He said a degree at the equator was 45 miles and the span of ocean between the Canary Islands and Japan only 2,765 miles—20 percent of the actual figure. If he had not providentially bumped into America, Columbus would—as the experts in Salamanca believed—have run out of food and fresh water long before he reached Japan. Columbus may have had courage and a "kindling eye" on his side; his opponents had science and reason on theirs.

Yet it was Washington Irving's version of history that became common knowledge, reprinted in 175 editions before 1900 and still appearing in textbooks and history books today. Why does the Flat Earth

Error remain so popular? Americans like to think that before we were discovered, all the world was sunk in darkness.

There is also the War Between Religion and Science to take into account. The groundwork for the war was laid in the 1850s by William Whewell of Cambridge University, who coined the word "scientist" to replace "natural philosopher." His *History of the Inductive Sciences*, which became a standard textbook, portrayed religion as inimical to science. He introduced two sources as proof that medieval Christians believed the earth was flat: the heretical Lactantius and the unread Cosmas Indicopleustes.

John W. Draper, a professor of medicine at New York University, expanded on Whewell's thesis. In 1860, in Britain, he presented a paper supporting Darwin's theory of evolution. He was attacked, viciously, by Bishop Samuel Wilberforce (son of William Wilberforce, the famous abolitionist). Remembering this painful experience as he wrote *The History of the Conflict Between Religion and Science* ten years later, Draper declared that science and religion were at war. Science stood for freedom and progress; religion meant superstition and repression.

Andrew Dickson White, the founder of Cornell University, put two and two together in 1896 in his *History of the Warfare of Science with Theology in Christendom*. "A few of the larger-minded fathers of the Church," White conceded, thought the earth was round, "but the majority of them took fright at once." He ignores Augustine, Isidore, and Bede (not to mention Gerbert and Abbo) and gives the majority view to Lactantius and Cosmas.

Over a hundred years later, the idea that medieval Christians like Gerbert thought the world was flat has not disappeared. It remains a weapon in the war between science and religion that defines modern America.

CHAPTER VIII

The Astrolabe

Ptolemy the astronomer was riding along on a donkey carrying a celestial sphere in his hand, according to an Arabic folktale. He dropped the sphere, it rolled under the donkey's hooves, and *squash*—the astrolabe was invented.

The Latin version of this story is more restrained: It leaves out the donkey. Ptolemy excelled in the study of stars, says a book from Gerbert's time. Among the instruments he invented was one that was "both useful for learners and a mighty miracle for those looking at it. . . . For the Wazzalcora was obtained by a divine mind; in Latin it means 'flat sphere,' which also, by another name, is the astrolabe."

Astrolabe means "star-holder." Harder to make but handier to carry than a celestial sphere (as the donkey anecdote shows), the astrolabe had more than a thousand uses—an Arabic astronomer in around 960 claimed it had precisely 1,760 uses. You could tell time by sun or stars, in terms of twenty-four hours of equal length or using the medieval clock, in which day and dark are each divided into twelve parts. You could find the celestial coordinates of the sun, moon, stars, or planets, and the terrestrial latitude and longitude of any town. You could calculate sine functions, tangents, declinations, and right ascensions. You could predict the time of an eclipse. When al-Khwarizmi and the astronomers of the House of Wisdom calculated the circumference of the earth, traipsing through the Iraqi desert at the behest of the caliph of Baghdad in 827, they used astrolabes to track the altitude of the sun.

Al-Khwarizmi's book on the astrolabe was known in Cordoba in Gerbert's time—no one can say how long it had been there. But in 978, Maslama of Madrid, the chief astronomer in al-Andalus, adapted al-Khwarizmi's star tables to the coordinates of Cordoba. To do so, he used al-Khwarizmi's book, as well as an Arabic copy of Ptolemy's *Planisphere*, which explains the math behind the astrolabe.

Parts of both books were translated into Latin before the year 1000, possibly at the monastery of Ripoll. The author wrote a rough version, in sloppy Latin and peppered with Arabic words; he added a great deal of explanatory material, along with a preface pointing out the astrological significance of the star of Bethlehem. He seems to have had an actual astrolabe in front of him, along with someone who knew Arabic well; this person identified the Arabic names of the various components and explained them to him in Latin.

At about the same time, someone in Catalonia made the first European astrolabe—a rather crude beginner's model, but proof the concept was understood. With the addition of this tool, medieval Latin astronomy suddenly became a truly mathematical science.

What part did Gerbert, the leading mathematician and astronomer of the West, play in this transition? Did Gerbert know, read, or even write that first Latin book on the astrolabe? Was he the scribe sitting beside the Arabic speaker at Ripoll? (Was his Latin *ever* sloppy?) Did he write one of the more elaborate revisions of the text that soon circulated through Western Europe? Experts have argued these points for a hundred years. The sources, admittedly, are none too clear. Yet it seems likely, at least, that Gerbert learned about this magical instrument while he was in Catalonia, and that he introduced it to his students at Reims.

William of Malmesbury—never too reliable—claimed that Gerbert "surpassed Ptolemy in knowledge of the astrolabe."

Michael Scot, who called Gerbert the finest necromancer in France, offered the astrolabe as evidence of his pact with the devil. Writing in the thirteenth century, he said Gerbert "borrowed" an astrolabe, conjured up his familiar demons, and forced them to explain how it was made, what it was good for, and how to work it. "He also made the demons teach him all astronomy. Afterwards the astrolabe came into the hands of many, and consequently there were many doctors of this art of various nations, regions, and times who compiled the books based on experiments." (If you substitute "Muslim" for "demon," this tale might be true.)

One of those "books based on experiments" is quite specific about Gerbert's relation to the astrolabe. After introducing the instrument as useful for measuring heights, lengths, and depths, the author adds, "It should be noted that Gerbert wrote a book on the astrolabe, which one can find in this volume in the second part, but it is rather confused; it does not teach how to construct the instrument, but only how to use one. Having read it, Berenger . . . asked his friend Hermann to provide him with a treatise on the construction of the astrolabe. It is in response to this demand that Hermann first composed this book, then he brought some order to that of Gerbert."

Berenger's complaint is that Gerbert told only how to *use* the astrolabe, not how to *make* it. It is the same criticism we could level at Gerbert's abacus book: He doesn't describe the instrument or provide any pictures.

Richer of Saint-Remy, who described Gerbert's abacus in such detail (the shieldmaker, the thousand counters of horn, the nine signs), does not mention the astrolabe at all; to some, that's sure proof Gerbert had never heard of one. Richer, unlike William (writing in the early 1100s) and Michael (in the 1200s), at least knew Gerbert personally.

But Thietmar of Merseburg, writing ten years after Gerbert's death, also knew him. In his history of the three Emperors Otto, Thietmar passes over the Scientist Pope in a few words. Yet what he points out is significant. In wording that strangely echoes the contemporary description of Ptolemy, Thietmar says Gerbert "was particularly skilled in discerning the movements of the stars and surpassed his contemporaries in his knowledge of various arts." By 1013, when Thietmar was writing, news of the astrolabe had spread throughout the West. Until the telescope came along in 1610, it would remain the most popular astronomical instrument. Gerbert could hardly have excelled in star-lore if he knew nothing about it.

Thietmar goes on. Gerbert made an *horologium* for the emperor, he says. "Astrolabe" is one possible translation: The word contains the root for "hour," and telling time was a popular use for an astrolabe. Thietmar might instead have meant a water-driven clock or *clepsydra*, a mechanical clock (which William of Malmesbury also claimed Gerbert had made), or some type of sundial. But he adds that Gerbert could position his *horologium* correctly only "after he had observed through a tube the star that sailors use for guidance," meaning the North Star. This detail fits one of Gerbert's celestial spheres or the instrument known as the *nocturlabe*, which consisted of a sighting tube, like a primitive telescope, surrounded by a graduated circle marked off in degrees. But it does not rule out the astrolabe.

Finally there is Gerbert's own aside, in his letter to Constantine, about using the abacus to multiply "actual numbers" in order to compare "the theoretical and actual measurement of the sky and of the earth." These numbers, he said, were "measurements determined by the inclination and erection of the geometrical radius." By "geometrical radius" he might have meant the sighting device of an astrolabe.

Given al-Khwarizmi's book—or a sample instrument—making an astrolabe would not have taxed Gerbert's mechanical abilities. It called for a metal workshop and a good hand at engraving, as had the brass tubes, horizons, armillary rings, and stands for his celestial spheres. It called for knowledge of climate circles and constellations. It required, as well, a mathematical theorem to accomplish, more precisely, what Ptolemy's donkey had done to the sphere.

This theorem was ancient. According to Bishop Synesius of Cyrene, who made an astrolabe of silver and gold in about A.D. 400, the idea of mapping a sphere onto a flat surface was "vaguely shadowed" by Hipparchus of Rhodes, born in 180 B.C. Hipparchus's so-called stereographic projection was recorded three hundred years later by Ptolemy in his *Planisphere.* Ptolemy did not invent the astrolabe, Synesius notes, but was content to have one as his "one useful possession, for the sixteen stars made it sufficient for the night clock." Synesius was a student of Hypatia, the first woman known to be a mathematician, and credits her help in making his deluxe astrolabe. The theory of the stereographic projection having been "neglected in the long intervening time" between Ptolemy and Hypatia, Synesius writes, "we worked it out and elaborated a treatise and studded it thickly with the necessary abundance and variety of theorems. Then we made haste to translate our conclusions into a material form, and finally executed a most fair image of the cosmic advance."

Where Synesius says "image of the cosmic advance," you could substitute "rotatable star map," "flat model of the starry sky," "analog computer," or "two-dimensional model of the universe that one can hold in one's hands"—all these have been used to describe the astrolabe.

Bishop Synesius's astrolabe, and all succeeding ones, had three basic parts: the base, or *mater* ("mother"); the latitude plates (one or several); and the "spider," known in Latin as the *rete*, or "net." The rete rotates around a pin, which also holds the sighting device, the arm or pointer called an *alidade*, to the back of the mater. The parts are locked in place by a "horse"—a wedge traditionally shaped like a horsehead—stuck through a hole in the pin. The astrolabe is carried by a ring at the top.

The spidery rete is the heavenly part of the instrument. The artistry of an astrolabe is in marking sixteen to fifty bright stars, in relation to each other and to the ecliptic, on a disc of brass, and then making that disc nearly transparent by cutting away every bit of metal that is not strictly necessary. What remains are thin brass circles and arcs. Delicate arrows and hooks, often labeled with the stars' names, jut from them to indicate the stars' locations: For the whole point is to be able to spin this map of stars above a latitude plate. One complete rotation of the star pointers over the latitude plate underneath it equals twenty-four hours of time.

A different plate is needed for the latitude of each location—Constantinople, Rome, Baghdad, Jerusalem, Cordoba—in which the owner hopes to use his astrolabe, so most medieval astrolabes were made with a stack of plates, each etched on both sides. Here's where Ptolemy's (or Hipparchus's) theory of stereographic projection comes into play. The great circles drawn on the surface of a three-dimensional celestial globe, or denoted by brass rings on an armillary sphere, are mapped onto a two-dimensional plate in a way that preserves the angles—two lines that cross at right angles on the globe will cross at right angles on the plate. Imagine you are standing at the south celestial pole and looking north, through the transparent earth, at the inner surface of the heavenly sphere, and that all the climate circles are visible: This is the point of view of the plate. The largest circle on the plate is the Tropic of Capricorn, the middle circle is the equator, and the smallest circle is the Tropic of Cancer. A series of arcs beginning at the horizon curve into full circles as they approach the zenith; these mark the degrees of altitude above the horizon (like lines of latitude on a terrestrial map). Large astrolabes can have an arc for each degree, whereas smaller ones make do with one arc for each 5 to 6 degrees. Another set of arcs, perpendicular to the horizon, emanates from the zenith point. These are the circles of azimuth, or direction measured around the horizon (like lines of longitude).

The latitude plates are held in place by the mater, a disc with a sturdy lip. On the back of the mater is engraved a calendar scale, showing days and months, and surrounded by a zodiac scale, telling what sign the sun is in each day. In the middle is a shadow square, for calculating the heights of towers and mountains using the geometrical principal of similar triangles. The rim of the mater is marked in 5-degree intervals to make a 360-degree altitude scale. To measure the altitude of the sun or a star, you dangle the instrument vertically from its ring and sight through the viewfinder of the alidade on the back; once the alidade is lined up with the star, you can read off the altitude from the scale on the astrolabe's rim.

To set the astrolabe to the correct time, take the altitude you just measured and find that star on the rete. Turn the rete until that star-pointer points to the correct altitude circle on the latitude plate. The result is a map of the sky at that very moment. Hours later, the star will be in a different place in the sky, at a different altitude above the horizon. To find out exactly how much time has passed, first note the current setting of the rete. Take the new altitude of the star, and rotate the rete until the star-pointer correctly indicates the star's new position in the sky. The amount you turned the rete—read with a time-pointer that projects from the rete over the 360-degree scale on the mater's rim—gives the time: 15 degrees equals one hour, with noon at the top and midnight at the bottom. Telling time by unequal hours was possible if an hour scale was etched onto the latitude plate. This convenient, pocket-sized marvel could thus accomplish every measurement Gerbert could make with his much bulkier celestial spheres.

If Gerbert did make an astrolabe, we have no record of it. The first mention of astrolabe-making in the West comes from Liege in about 1025, in the series of letters already mentioned between Rodolf of Liege and Ragimbold of Koln. The two friends were in the midst of their investigation of angles and triangles—their letters demonstrating the spread of Gerbert's experimental approach to geometry. Now Rodolf wrote to his friend, "I would have sent the astrolabe for you to judge it; but it serves us as an exemplar to construct another. If you wish to know about it, please be so good as to come to Mass at Saint-Lambert's. I trust you will not regret it."

What did the astrolabe Rodolf brought to Mass look like? We cannot know: No astrolabes exist that were made near Liege or anywhere in the north before 1540. In fact, of the 150 medieval European astrolabes extant, only two can be dated to before 1200.

Because of the precession of the equinoxes, an astrolabe is only accurate for about a hundred years. (This concept, that the equinoxes appear to be slowly moving westward in relation to the fixed stars of the zodiac, was first pointed out by Hipparchus; al-Battani wrote a treatise on it in the tenth century. We now know it occurs because the earth's axis wobbles.) Unless it is associated with someone famous, there's no reason to keep an out-of-date astrolabe. The brass is melted down and, especially in wartime, reused. As a result, most extant medieval astrolabes are "one-off" pieces—the only example known to have come from a single instrument maker or workshop. Yet most are not beginners' work. The Sloane astrolabe at the British Museum is the first English-made astrolabe, dating to about 1300. Its mind-boggling complexity suggests that English instrument makers had been producing astrolabes for hundreds of years. The oldest German astrolabe still extant is equally complicated.

Gerbert's name is linked to two astrolabes that do still exist. In Florence is a brass astrolabe long said to have been Pope Sylvester's. It was supposedly made for Gerbert in 990 or 1000 or 1002, in France or maybe in Cairo. Its later owners, the Medici family, however, kept it in a leather case along with an inscription on parchment saying it was made for King Alfonso X of Spain in 1252. A careful examination of the instrument shows it was made to be used in Baghdad. The distinctive form of Arabic *kufi* script used in the star names and other inscriptions dates it to the tenth century—which means Gerbert (and later King Alfonso) could have owned it, but not used it (unless they traveled south to the latitude of Baghdad, or owned an additional latitude plate). Much later, probably in the nineteenth century, an antique dealer etched on Latin lettering for the months, presumably to make it look as if the astrolabe were made in Spain. His mistake was giving the length of February as 28¼ days—he forgot that tenth-century fractions were base twelve; to people of Gerbert's time, "¼" would have been unreadable. But the astrolabe itself is a well-made scientific instrument.

The other astrolabe that Gerbert could have seen or used or owned is not well made. Known as the Destombes astrolabe, it is brass, the size of an open hand. Now in Paris, it was made in Catalonia in the tenth century, at about the same time (or before) the earliest Latin astrolabe texts were being written there. The oldest European astrolabe still extant, this crude, seemingly unfinished instrument, clearly fashioned by a beginner, marks the passage of Islamic science to the Christian West before the year 1000.

The Destombes astrolabe is the first known astrolabe with inscriptions in Latin. A beginner's model, it was made in tenth-century Catalonia by someone who was, most likely, copying from another astrolabe, or, possibly, from an illustration of an astrolabe.

Curiously, the Destombes astrolabe is unlike any of the sixteen Arabic astrolabes remaining from tenth- or eleventh-century al-Andalus. The rete carries eighteen stars (none of them named) and is so clumsy it seems to have been copied from another astrolabe (Rodolf's plan) or even from a drawing of an astrolabe. But it bears no resemblance to any Arabic rete. Six of the star-pointers are too long: To accurately mark their stars, they should have been curled up at the tips; this was never done. The placement of the bright star Aldebaran is seriously off: It marks the path Aldebaran took through the sky in Ptolemy's time; for this one star only, the precession of the equinoxes was not taken into account. The other eleven stars are correctly placed, matching the data in the star tables made by Maslama of Madrid in 978.

The instrument can also be dated to before 986—no one can say how much before—based on an inscription on one of the latitude plates. (If one plate is more detailed than the others, you can assume the instrument maker favored that latitude—the astrolabe was meant to be used there, or the maker was working there.) The inscription reads, in good tenth-century Catalan script: *Roma et Francia*. Working backward from the engraved circles, geometry proves the plate was designed for the latitude 41°30', which is close to that of Rome (41°53'). It is also within a few minutes of the latitude of Barcelona, the bordertown of "Francia."

Gerbert's letters to and about Spain reveal the significance of this inscription. As Gerbert shows, until 986 Catalonia was part of the kingdom of France; after that, the astrolabe maker would have written *Roma et Burcinona*.

The story of Catalan independence, as Gerbert tells it, begins in 976 with the death of the scholarly caliph al-Hakam, to whom Gerbert's friend Miro Bonfill had been sent as ambassador. Al-Hakam was succeeded by his ten-year-old son, who was overshadowed by the warleader al-Mansur the Victorious. Al-Mansur propped up his power by warring on Christians. In July 985, he sacked Barcelona. Gerbert's friend Count Borrell called for aid to his overlord, the king of France. There was no reply.

The king of France died in early 986 and was succeeded by his young son, Louis V, known as "Louis Do-Nothing." On Borrell's behalf, Abbot Gerald of Aurillac wrote to Gerbert at Reims to ask "what sort of person" Louis was and whether he was "about to conduct the armies of the French as aid to Borrell." Replied Gerbert, "It is unnecessary to ask the first question of us, because, as Sallust says: 'It is necessary that all men who take counsel about doubtful matters should be free from wrath, animosity, and compassion.'" As we will see, Louis had just accused Archbishop Adalbero of Reims of treason. "Considering the other question according to its nature," Gerbert continued, ". . . our opinion seems rather to incline to the negative."

Borrell petitioned France again in 987. This time he was answered by a letter Gerbert wrote as secretary to Hugh Capet, the third king to rule France within that span of two years. "If you wish to maintain the fidelity so often offered through intermediaries to our predecessors," said King Hugh, "hasten to us with a few soldiers in order to confirm the fidelity already promised, and to point out the necessary roads to our army." Borrell never left Catalonia, the king's army never left France, and historians date Catalan independence from 986.

We do not know if the Destombes astrolabe, or one like it, came north from Catalonia before 1025, when an astrolabe landed in Rodolf of Liege's hands. But news of this magical instrument had. Though Liege is a long way (nearly 900 miles) from Barcelona, Rodolf had no need to tell Ragimbold what an astrolabe was: The instrument had been integrated into the quadrivium long before, probably by Gerbert.

Fulbert of Chartres, for example, was master of the Chartres cathedral school by 1004. He wrote several mnemonic poems to help his students understand scientific concepts. One is on weights and measures, another on the calendar; a third, strangely, is mostly in Arabic:

Abdebaran Tauro, Geminis Menkeque Rigelque,
Frons et Calbalazet prestant insigne Leoni;
Scorpie, Galbalagrab tua sit, Capricornie, Deneb,
Tu, Batanalhaut, Piscibus es satis una duobus.

In English it reads, "Aldebaran stands out in Taurus, Menke and Rigel in Gemini, / and Frons and bright Calbalazet in Leo; / Scorpio, you have Galbalagrab, and you, Capricorn, Deneb, / You, Batanalhaut, are alone enough for Pisces." The Arabic words, of course, are the names of stars, some of the brightest in the sky. Though, like Gerbert, Fulbert never uses the term "astrolabe" (or Wazzalcora, for that matter), anyone familiar with the instrument would recognize these stars: They are the stars on the rete of the astrolabe that indicate the climate circles.

Fulbert also left a notebook in which he defined twenty-eight Arabic words used in connection with the astrolabe and supplied a chart showing the twelve signs of the zodiac in one column and the name of an astrolabe star (if there was one for that zodiac sign) in a second column. The Arabic terms, and most of the Latin translations, come from the first Latin book on the astrolabe, the rough and sloppy one compiled in Catalonia. But some of Fulbert's explanations expand on the Catalan book, showing that he had information from elsewhere, as well as a good grasp of how to use the actual instrument.

Fulbert, who became bishop of Chartres in 1006 and died in 1028, may have been Gerbert's student; the sources, again, are unclear. One theory of how he learned about the astrolabe, then, is that Gerbert wrote the first Latin book on the subject at Ripoll—or simply found it there and copied it—and brought news of the astrolabe north to Reims, where Fulbert was a student. Yet as an adolescent, arriving in Catalonia in 967, Gerbert was already known for his beautiful Latin style. The Ripoll book, besides being written in poor Latin, retains many Arabic terms; if Gerbert knew these Arabic words, why don't they show up in any of his other writings (as they do in Fulbert's)? Gerbert may have owned a copy of this first Latin book on the astrolabe, but he is unlikely to have written it.

Another theory concerning this book names Lobet of Barcelona as the author. Gerbert wrote to Lobet in 984 requesting *"De astrologia, translated by you."* He received it and subsequently revised it, the idea goes—just not very well, allowing later writers to say Gerbert's astrolabe book was "confused" and needed the help of Berenger's friend Hermann to bring some order to it.

Yet *"De astrologia"* is vague: *On the Study of the Stars* might refer to this book on the *astrolabe*. But it might better refer to a book on *astrology*. The oldest Latin manuscript to contain Arabic words is, in fact, a book on astrology. Now in Paris, it is a rather thin little book, with a newish leather binding. The parchment is off-white with faded brown letters, undistinguished. Capital letters, meant to have been painted in red, are missing. Abruptly, midway, the writing becomes much smaller, to save space. It was made in Limoges, 400 miles north of Ripoll, between 978 and 1000. Parts were copied from an earlier manuscript, now lost, that came from Catalonia; other parts came from the monastery of Fleury, including a work by Abbo on cosmology. Most of it, however, is about fortune-telling, not what we would call the science of the stars. The collection of astrology texts it contains, named the *Alchandreana* after one of its purported sources, Alchandreus, can be found in nearly a hundred manuscripts, fifteen from before the year 1200.

The Arabic words appear on a page of diagrams. Two have not been deciphered. One clearly shows the twenty-eight lunar mansions, with a drawing of the stars in each. The star names are given in Arabic. Beneath the illustration starts a text identified in 2007 as having been written by Gerbert's friend Miro Bonfill. Based on the playful, pun-filled style, two of the other astrological treatises in the *Alchandreana* can also be attributed to Miro. The three others, which begin with *De astrologia*, have been ascribed to Lobet of Barcelona—with just as much (or as little) evidence as the book on the astrolabe from Ripoll is thought to be his work. We have a book about stars and Gerbert's letter asking Lobet for a book about stars.

The *Alchandreana* is three times as long as the Ripoll book on the astrolabe. Based on Arabic and Jewish sources, and using Hebrew, Latin, and Arabic letters, it is well-organized and clear, and it provides dozens of sample computations for predicting the outcome of an illness; the character of a child; the success of a journey, a marriage, or a battle; the site of buried treasure; or the identity of a thief. Fifteen chapters tell how the *phisici* forecast the weather.

Gerbert, a close friend of Miro Bonfill, most likely knew this book and practiced astrology himself—which was not considered pseudo-science in the tenth century. A knowledge of astrology would explain why Gerbert was so welcomed by the pope and the emperor, when he came to Rome from Spain in 970, as a master of *mathesis*—a word more commonly used for fortune-telling than for mathematics. According to Richer of Saint-Remy, *mathesis* was at the time unknown in Italy.

Astrology also explains Gerbert's later reputation as a necromancer. When William of Malmesbury says Gerbert "surpassed Ptolemy in knowledge of the astrolabe," he follows it immediately by saying he also surpassed "Alchandreus in that of the relative positions of the stars, and Julius Firmicus in judicial astrology." He then tells a fanciful tale of Gerbert using his knowledge to find buried treasure—something Miro describes how to do in the *Alchandreana*.

Nor was William the first to accuse Gerbert of being a wizard. A poem by Ascelin of Laon, the nephew of Adalbero of Reims, names King Robert's tutor "Nectanabo." Nectanabo is the name of the wizard in the legend of Alexander the Great, and Alexander is one of the authorities on astrology mentioned in the *Alchandreana*. Everyone hearing Ascelin's poem would have known Gerbert had in fact been the king's tutor and that his "wizardry" included telling horoscopes.

Finally, Gerbert uses astrological lore in his own writing. He dates the death of a duke to June 17, 983, by saying "the Sun found itself in the house of Mercury." To an astrologer, June 17 of that year was the day the Sun transited the last degree of Gemini, which is the "house" of Mercury. The theory of the houses of the planets is explained in the *Alchandreana*.

And then there is Gerbert's complaint about the floods on the way to Trier: "The hope for better weather has been shattered by the *phisicis*," he writes. He may have meant himself.

Jotted on the last page of the manuscript is a horoscope that a monk worked out for himself in 1014. The monk was Ademar of Chabannes, who described Gerbert in his chronicle as seeking knowledge among the Arabs in Cordoba. The method Ademar used to compute his horoscope—the method taught throughout the *Alchandreana*—does not depend on a study of the stars, but on affixing numbers to a person's name. Ademar first translated his own name and that of his mother into Hebrew. Then, using a code in which the letters of the Hebrew alphabet stand for numbers, he found the numerical values of the names and began his calculations—for to be a fortune-teller meant being a very good mathematician. Depending on the question you were asking, you may have divided by the number of years since the creation of the world. If Saturn figured in your horoscope, you would have divided that number by 30 (the number of Earth years that makes one Saturn year); likewise, by 12 for Jupiter, or by 1½ for Mars. For Venus, you used the number of days, not years, dividing by 300.

An abacus would come in handy, and many copies of this astrology book are bound with treatises on the abacus. They are also found alongside treatises on the astrolabe: To fully use the method, you needed to be able to tell the exact time.

Two manuscripts, one from Lorraine and one from Bavaria, preserve short bits of the *Alchandreana* that are intelligent and well written— and appear alongside Gerbert's own works on mathematics. These bits could be all that is left of Gerbert's own textbook on astrology, based on Lobet's work.

If Lobet of Barcelona's *De astrologia* was about astrology, how did Fulbert of Chartres and Rodolf of Liege learn about the astrolabe? Was Gerbert the link between Catalonia and the north? It's still possible. Gerbert was a teacher, not a writer. All of his known scientific treatises were written at the request of a student. Having learned about the astrolabe, his first impulse would not have been to write a book. Instead, he would have incorporated the astrolabe into his teaching of astronomy. His students would have spread the knowledge farther.

Fulbert, Rodolf, and Ragimbold were all members of the same scientific network. Rodolf and Ragimbold conferred with Fulbert on mathematical questions; as Ragimbold wrote, "I passed by Chartres, and Lord Fulbert, bishop of the place, demonstrated to me your same figure, with an exposition of our first question concerning the triangle; and, after many conferences, he agreed with our opinion." And Fulbert knew Gerbert's scientific works: A tenth-century catalog shows that the Chartres library held copies of Gerbert's letters to Constantine on the abacus and the celestial sphere.

The key, in fact, is Gerbert's student Constantine, whom he called "sweet solace of my labors." Without Constantine, we would know very little about Gerbert. Under the pen name "Stabilis," as we have seen, Constantine preserved and copied Gerbert's letter collection. Constantine induced Gerbert to write about the abacus—an exercise he found "nearly impossible"—and shared Gerbert's treatise with other scholars, like Fulbert. Constantine quizzed his former teacher on those baffling problems in Boethian number theory. He requested instructions on making a celestial sphere. The sphere's star-viewing tubes "differ from

organ pipes by being all equal in size," Gerbert notes, suggesting he and
Constantine had also discussed his treatise on organ pipes, which would
have greatly interested Constantine: He was known as "a remarkable
musician." Gerbert recommended that the monks of Aurillac consult
him about "the learning of music and the playing of organs," adding, "I
will see to it that what I am unable to finish myself will be completed
by Constantine."

Not surprisingly, Constantine also collected information on the astrolabe. He seems to have had the "confused" treatise later attributed to
Gerbert—the one that "does not teach how to construct the instrument, but only how to use one." He may have received it from Gerbert. Or maybe not. We have no evidence. We know only that he
sought to supply what was missing and was successful. "Ascelin the German, citizen of the city of Augsburg, to Stabilis of Orleans, monk of
Micy, greetings!," begins one astrolabe manuscript. "Concerning how
much effect firm friendship has in getting things done, . . . I have decided that the effort of my whole talent should always reply kindly to
the wishes of friends." After complimenting his "beloved" friend for
"being of unsullied honesty" and "by your own name and at the same
time truly by your character . . . being 'Stable, Constant,'" and therefore
"not unequal of the laws of friendship," Ascelin asks him to "accept,
then, this work that you desire, worked out for constructing the instrument of the astrolabe, not completely, but diligently according to
the small measure of my intelligence."

Ascelin's book is also confusing. It has no illustrations. It is very difficult to follow his directions without a series of drawings, breaking them
down step-by-step, as most later astrolabe treatises do; or without an actual astrolabe to look at. Perhaps Constantine had one. Astrolabes may
have been more common in the late tenth century than we think. Certainly the knowledge of how to build them had reached Augsburg in Germany, 800 miles north of Barcelona, while Constantine was at Micy.

Knowing when that might have been requires some sleuthing among
saints' lives and in church archives. We learn that Abbo of Fleury so
disliked Gerbert's beloved Constantine that he refused even to hear his

music played in the church at Fleury. Only after Abbo died, says the *Life of Gauzlin*, the next abbot, was there "first performed at Fleury the story of the arrival of Saint Benedict which Constantine—a man raised in that place but later given the honor of the abbacy of Micy by Arnulf, the bishop of Orleans—had written." No surprise, then, if Constantine left Fleury as soon as he could after Abbo became abbot in December 988. He was clearly one of the monks who "perversely resisted" Abbo's election. In fact, Bishop Arnulf, Gerbert, and the archbishop of Reims had all lobbied for Constantine to be named abbot of Fleury himself.

Instead, Constantine went to Micy, a smaller monastery outside Orleans, about 30 miles from Fleury. A few years later, Bishop Arnulf named him abbot of that monastery. When? There are two possibilities. A Constantine became abbot of Micy in 1011, but this cannot be our Constantine—as most historians of science have thought—because Bishop Arnulf of Orleans died in 1003.

The first Abbot Constantine of Micy—our Constantine—was appointed before 994, for in that year the count of Aquitaine kicked him out. This news comes from a curious tale of miracle-working saints' relics owned by the monastery of Nouaille, over 100 miles away from Micy. The story was written by a monk named Letaudus of Micy. He was asked to write the story by Abbot Constantine of Nouaille, whom he had known for several years, he says, for Constantine had formerly been the abbot of Micy. The first church document to name an Abbot Constantine of Nouaille is dated August 994; he died there in 1014.

Constantine may have considered himself still the rightful abbot of Micy until his death, for by moving him to the less-valuable monastery of Nouaille, the count of Aquitaine had broken a solemn oath, sworn over holy relics, not to meddle in Nouaille's affairs. To Constantine, the new abbot of Micy was a usurper. To enshrine his dissatisfaction, he commissioned the story from Letaudus. Its moral: The saints will have their revenge on oathbreakers.

But Ascelin, the German, directed his book on the astrolabe to a "monk of Micy," not the abbot, thus dating it between December 988

and August 994. Gerbert was still at Reims, but while the cult of friendship, around which Gerbert organized his school, infuses Ascelin's letter, we can identify him only as Constantine's friend. The Ascelin we
know to have been Gerbert's student, the nephew of Archbishop Adalbero (and the poet who called Gerbert "Nectanabo" the wizard), does
not seem to have had any connection with Augsburg, Germany. Nor
can we say what connection Augsburg had with Catalonia, where the
astrolabe texts—and instruments—originated.

But one connection is clear: Gerbert and Constantine were the closest of friends. If Constantine was studying this magical instrument between 988 and 994, Gerbert knew of it. In December 988, upon Abbo's
election as abbot of Fleury, Gerbert invited his friend to visit Reims.
Perhaps they discussed the astrolabe then. Perhaps Gerbert gave him
the "confused" treatise or simply explained what he had learned in Catalonia years ago. Why Constantine returned to the Orleans area and
worked on his astrolabe book there, without Gerbert's further help, will
become obvious as we explore the nonscientific side of Gerbert's life.
For in January 989, his mentor Adalbero of Reims died, and Gerbert's
world slipped "into primordial chaos." He could not answer Constantine's requests—assuming he had asked—for more information on the
astrolabe. He could not even finish the celestial sphere he had begun
making for Remi of Trier. As he wrote to Remi, "Hence, endure the de-
lays imposed by necessity, awaiting more opportune times in which we
can revive the studies, now already ceasing for us."

The leading mathematician and astronomer of his day was about to
become embroiled in some very messy politics: a coup and civil war in
France, revolts and assassinations in Rome, the kidnapping of the heir
to the imperial throne. Gerbert was at the center of the tumult, and his
mastery of the abacus, Arabic numerals, geometry, acoustics, celestial
spheres, the astrolabe, and astrology helped him not at all—except to
bring him to the attention of counts, kings, and emperors.

Plate 1 The Majesty of Saint Foy, from the Abbey Sainte Foy at the cathedral of Conques, France. This golden, jewel-encrusted reliquary, made in the tenth century from recycled Roman statuary and jewelry, holds the bones of a thirteen-year-old girl martyred six hundred years previously. It was so popular that the abbot of Gerbert's monastery in Aurillac had a similar majesty made of its founder, Saint Gerald.

Plate 2 The dedicatory page from Saint Bernward's evangelary. Making books was a monk's sacred duty. Here, Gerbert's contemporary, Bernward of Hildesheim, places a book on the altar, symbolically presenting it to the Virgin Mary (who is shown on the facing page in the manuscript).

Plate 3 This illustration from a thirteenth-century Turkish manuscript shows Aristotle teaching his students how to use an astrolabe, the most popular astronomical instrument in the Middle Ages. Gerbert seems to have learned how to make and use an astrolabe while he was in Catalonia, and must have taught it at Reims. The first Latin text on the astrolabe was most likely written by his student, Constantine of Fleury.

Plate 4 The arqueta presented to Gerbert's patron, Count Borrell of Barcelona, by the Islamic caliph of Cordoba between 961 and 976. The box of wood, gold-plated with silver decorations and pearl accents, was signed by a Jewish artist and is now at Girona cathedral, affording proof, not only of exchanges between the Christian and Muslim kingdoms, but of the inclusion of Jews in the culture.

S S S S

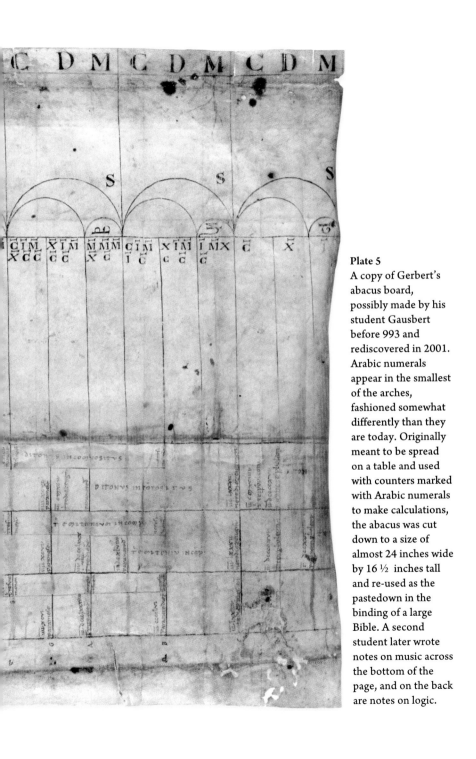

Plate 5
A copy of Gerbert's abacus board, possibly made by his student Gausbert before 993 and rediscovered in 2001. Arabic numerals appear in the smallest of the arches, fashioned somewhat differently than they are today. Originally meant to be spread on a table and used with counters marked with Arabic numerals to make calculations, the abacus was cut down to a size of almost 24 inches wide by 16 ½ inches tall and re-used as the pastedown in the binding of a large Bible. A second student later wrote notes on music across the bottom of the page, and on the back are notes on logic.

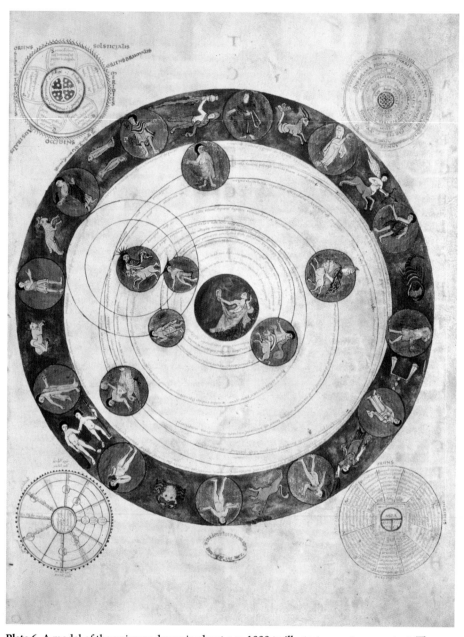

Plate 6 A model of the universe drawn in about A.D. 1000 to illustrate an astronomy text. The earth, sun, moon, and five planets are surrounded by the zodiac. Note that while Earth is at the center of the universe, only the moon has a circular orbit. The orbits of the sun, Mars, Jupiter, and Saturn around Earth are eccentric, while Mercury and Venus circle the sun.

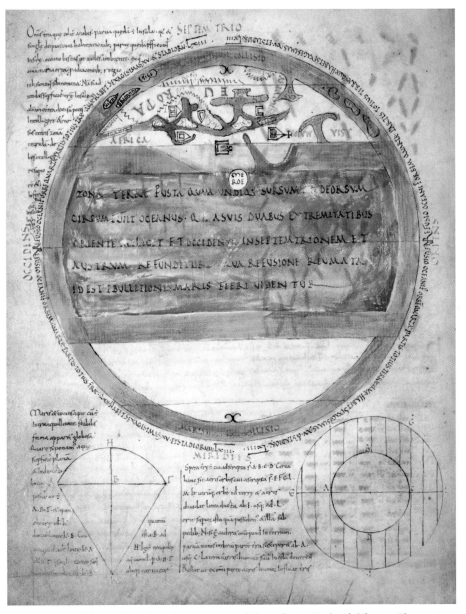

Plate 7   A map of the world made at the monastery of Fleury during Gerbert's lifetime. The captions explain that only one hemisphere of the globe is being shown, and refer the student to Eratosthenes' method for measuring the circumference of the earth. The map also predicts that there is inhabitable land in the southern hemisphere matching that known in the northern hemisphere.

Plate 8 The coronation of Otto III as Holy Roman Emperor in 996, from a Gospel book made for the emperor at the monastery of Reichenau. The older cleric on the left—with his hand on the throne—could be Gerbert, who was Otto's secretary and chief counselor at the time.

PART THREE

FROM ABBOT
TO POPE

*This event will accomplish what I have always
wished, have always hoped for, by making me his
inseparable companion so that we can guide the
exalted empire for him. What, therefore, could
be sweeter?*

GERBERT OF AURILLAC, 997

CHAPTER IX

The Abbot of Bobbio

The death of Archbishop Adalbero of Reims in 989 ended Gerbert's career as a schoolmaster. But his scientific studies had been "interrupted for a time," as he put it, once before—"though always kept in mind," he added.

Nine years previously, in 980, the fame of Gerbert's school at Reims had aroused the envy of Otric, the schoolmaster at Magdeburg in Germany. A favorite of Emperor Otto II, Otric was acknowledged as the supreme intellectual of the Holy Roman Empire. He sent a student—a monk who "seemed capable of the mission"—to infiltrate Gerbert's school.

Otric's spy returned to Magdeburg with scandalous news. Otric warned the twenty-five-year-old emperor that this upstart "knew nothing about philosophy" and should not be permitted to teach. Otto II was skeptical. He had become emperor upon his father's death in 973, a year after Gerbert had left court, and remembered his former Latin tutor fondly. He summoned Gerbert to court, not explaining why.

Thanks to Otric's envy, Gerbert's life changed: He won fame and fortune and imperial favor. Otto named him abbot of Bobbio, the Italian monastery renowned for its library—where he quickly found himself slandered and in danger of his life. Though he escaped and took refuge with Adalbero at Reims after three years, he never again could devote himself wholely to science: He was distracted by imperial dreams. From 980 to 989, his time was increasingly taken up by politics. From 989 until his death in 1003, he was totally embroiled in the games of kings. Who knows how science might have flourished if he had remained a simple schoolmaster? We do not know if Gerbert had learned about the astrolabe by 980, or written his treatise on organ pipes. What else might Constantine have convinced him to write if he had stayed out of politics? Yet if he had not come under the emperor's eye in 980, Gerbert would not have become pope—and we might never have heard of him.

The emperor's summons was for Christmas at Ravenna; so, late in the year, Gerbert accompanied Archbishop Adalbero south. At Pavia, Italy, they caught up with Otto and the young Byzantine princess who was his wife. Empress Theophanu was the one to impress; her wishes, more than anyone's, would shape Gerbert's future.

The imperial party traveled by barge down the Po River, past orchards and pine forests, to Ravenna. To Gerbert the city would have seemed shockingly Byzantine. Its great stone basilicas were lavishly decorated with mosaics. Knots and laces, diamonds and circles, swirls of figure eights and intricate images—a ring of dancers, a shepherd and his flock—adorned the floors in black, white, and red cut stones. On the walls and ceilings the colors were vibrant: gold and sparkling pearl, turquoise and peacock, red, magenta, orange, pink. The panels, made of twinkling bits of glass and shell, were alive with doves and deer, leaves and flowers, sparkling stars, the Lamb of God in every possible setting, Cain and Abel, Abraham and Isaac, the Burning Bush, Daniel in the lion's den, the Madonna and Child, the Three Wise Men, the baptism of Christ (a heretical version that showed him naked and fully human), the resurrection of Lazarus, the Last Supper, the four evangelists (represented by their traditional symbols: angel, lion, eagle, and bull, and the four Gospels in a tidy bookcase), the twelve apostles, fifty-five saints, the cities of Bethlehem and Jerusalem, and Christ Triumphant, seated on a blue globe. In the great basilica of San Vitale were portraits of the artists' patrons: Emperor Justinian, clad in purple, who ruled from 527 to 565, and Empress Theodora, as stern and lovely as a goddess, leading a train of court ladies with large, dark, deep-set eyes. Otto's Byzantine wife would have seen herself in their faces—and admired their lush, shimmering gowns.

Theophanu was not the "purple-born" princess Otto the Great had wanted for his son. According to Thietmar of Merseburg, who wrote his *Chronicon* between 1013 and 1018, Theophanu was not born in the purple room of the imperial palace as the daughter of a reigning emperor; Thietmar implied that the Byzantines had swapped the promised princess for a girl of lesser rank. But, if not born to the purple, Theophanu faked it well. In the Byzantine court, the emperor and his men were always balanced by the empress and her women; she held female rites to match his male rites, and joint ceremonies could not take place if the empress was missing. Theophanu insisted upon the same treatment in the West. Unlike most Western queens, she accompanied her husband on all his travels. Medallions, ivory carvings, and manuscripts all presented the two as equals.

She had been crowned on her wedding day, aged twelve. She did not have her first child, Adelaide, until five years later; then came Matilda and Sophie and, in 980, Otto III and a twin sister, who died. Knowing only Greek when she arrived, Theophanu learned Latin and the local languages so well that she was known as *ingenio facundam*, "a genius of eloquence" (or, by a not-so-kind monk, unpleasantly talkative). Another praised her moderation and good manners (adding that "this was exceptional for a Greek"). Her mother-in-law, on the other hand, referred to her, dismissively, as "that Greek woman."

From Byzantium she had brought "a splendid entourage and magnificent gifts." We don't know if any Greek books were among her treasures (though she did choose a Greek-speaking tutor for her son). What later ages recalled were the chess pieces and perfume bottles cut from precious stones. The ivory carvings in an oddly realistic style. The bold wall hangings and robes of silk—one pattern bore golden lions ranked on a plane of royal purple. Theophanu wore a necklace made like a sparkling net of gold, with six rows of nine jewels across; each jewel had another dangling from it.

After her death a nun saw her in a vision, weeping and begging for prayers. Theophanu had been damned for having brought a taste for silks and jewels to the women of the West—damned like another Byzantine bride who had offended the Almighty with her vanity. According to an eleventh-century churchman, "such was the luxury of her habits" that she did not deign "to touch her food with her fingers, but would command her eunuchs to cut it up into small pieces, which she would impale on a certain golden instrument with two prongs and thus carry to her mouth." It is the first mention, in the Christian West, of a fork.

Praised or damned, Theophanu attracted notice. Her husband did not. Otto II was generally dismissed as not the man his father was. Says Thietmar, "As a young man he was noted for his outstanding physical strength and, as such, initially tended towards recklessness." After "enduring much criticism," he learned to restrain himself and to listen to his elders. "Thereafter, he comported himself more nobly." Possibly his lessons with Gerbert in Socratic disputation helped him here.

He may also have picked up from Gerbert a love of books. His visits to monasteries were universally feared. At Saint-Gall, the *Life of Meinwerk* records, Otto asked to see the library's finest books. "The abbot hesitated, knowing well the king would take some. Unable to refuse, the abbot showed Otto the books and Otto did take some. Only after considerable correspondence did he return a few."

In addition to books, Otto enjoyed a rousing scholarly debate. To stage one between Otric and Gerbert, he thought, would wonderfully liven up the usual Christmas festivities. To Ravenna in December 980, therefore, he summoned a great number of schoolmasters and scholars. They gathered in his palace under the glittering mosaics a few days after Christmas. The emperor and empress mounted their thrones, and Otto called Gerbert and Otric before him. He still had not told Gerbert why he had been summoned. "He hoped that if Gerbert were attacked without warning, he would pour more passion and fire into contradicting his adversary," writes Richer of Saint-Remy.

According to one of Gerbert's friends who attended the debate, the emperor set the stage with a few flowery words about how learning ennobles the spirit. Then Otric stepped forward and laid out his complaint: Gerbert was teaching physics as if it were a subdiscipline of mathematics, whereas every educated person knew that physics and mathematics were two different and equal fields of study. Gerbert knew nothing about how knowledge was organized.

This was no "how many angels can dance on the head of a pin" exhibition, but a serious attempt to categorize the sciences. Professors today hold the same debates: Twenty-first-century academics are at odds over whether archaeology is a type of history or should be taught as a science. Yet according to Richer, Otto and his court considered this esoteric scholarly joust to be fine Christmastime entertainment. The battle of wits went on all day, with point and counterpoint, questions from emperor and audience, and lengthy citations from Plato, Boethius, and other authorities.

Physics and mathematics were, *of course*, two separate fields of study, Gerbert responded. Otric's spy had misunderstood. Math and physics (which included medicine), together with theology, formed the theoretical side of philosophy. The practical side, as Boethius had written, included ethics, politics, and economics.

Having lost that point, Otric seized on an aside and asked, "What is the purpose of philosophy?"

"To allow us to understand all things human and divine."

"Why do you take so many words when one is enough?" quibbled Otric (though he did not say which word he had in mind).

Answered Gerbert, "Not every answer can be reduced to one word." How do you explain what creates a shadow in one word? The cause of a shadow is a body placed in front of a light. If you say, "The cause of a shadow is a body," your definition is too general. If you say "a body in front," the definition is worthless, for many bodies can be placed in front of other things without causing a shadow.

Beaten again, Otric changed the subject: Which is more comprehensive, the rational or the mortal? Gerbert seized on the question with delight. His sentences "flowed on in abundance" until finally the emperor called a halt and proclaimed Gerbert the victor.

He was "covered in glory." Otto granted him "rich presents" and appointed him abbot of the monastery of Saint-Columban in Bobbio. The post gave Gerbert the rank of a count and required him to swear fealty to the emperor and empress as his liege lords. Gerbert prided himself on never breaking that oath, or making another. From then on, his future was shaped by his duties to the empire.

Otric, however, lost favor at court. A few months later, the emperor vetoed his election as archbishop of Magdeburg and chose another. Otric took sick and died.

After his dazzling performance, the new abbot of Bobbio accompanied the emperor and Archbishop Adalbero to Rome. There, at an Easter synod, Gerbert saw his Catalan friend Miro Bonfill, bishop of Girona.

Leaving Rome after the synod to return to their churches, Adalbero and Miro could have kept company with Gerbert north to Piacenza—where Count Gerald the Good had once bribed the ferryman to take him across the Po River. There Gerbert left the group and turned west toward Bobbio, a day's journey away. The glowing green fields gave way to hills, a blue haze hinted of higher mountains beyond. A castle on a hill—ramparts and a round tower stacked up from yellow stone—looked down on the river Trebbia, wide and low and stony. The road wound into the narrow river valley: The tops of the hills turned to bare rock; the river now sported rapids. Crossing a ridge of rugged hills, he finally reached Bobbio.

The town huddled beside the Trebbia, between two round knolls. Its humpbacked Roman bridge, the ten gray-stone arches each a different size, spanned a rocky narrows. The same gray stone was laid up into a drystone wall ringing the town. It paved the narrow streets and lined the channels by which a hilltop cascade was steered downhill to a water-wheel. The monastery, the cathedral, the blocky castle on the hill—all were made of the same gray stone. Bobbio was a fortress of a town—organized, protected, safe, strong, a center of administration in this untamed corner of Italy.

At first Gerbert reveled in the immense library of 690 books. He found there Boethius's *On Astrology*, "some beautiful figures of geometry," and other volumes "no less worthy of being admired," he wrote. But Bobbio was more than its books: It was the largest landowner in northern Italy. Its holdings stretched from Genoa on the Mediterranean north to the lake region, south into Tuscany, and east along the valley of the Po. Gerbert's task here was not to enlarge minds. He was an administrator. As abbot and count, he was required to provide soldiers and support for Otto's wars. He was not up to the task.

Gerbert was bitterly aware of his failure. Writing to a former student, he riffed ironically on a phrase of Cicero's that Otto had quoted in the diploma appointing him abbot: "*In proportion to the greatness of my mind*, my lord has enriched me with very extensive properties. For what part of Italy does not contain blessed Columban's possessions? This, indeed, our Caesar's generous beneficence provided. But fortune decreed otherwise. For, *in proportion to the greatness of my mind*, she has honored me with enemies everywhere."

He was lost in a world of intrigue. It was at Bobbio that Gerbert first began saving copies of his correspondence, to protect himself from the "foxes" who crawled the imperial palace at Pavia, flattering Otto, whispering and plotting, shamelessly slandering the new abbot as a stud horse, "as if I had a wife and children, because of the part of my household brought from France." Incidentally, this comment is the closest we come to knowing if Gerbert had a sex life: His enemies suspected it and spread rumors to that effect. Bobbio could not pay its dues to the emperor, they intimated, because of the luxury in which the new abbot and his family lived.

Gerbert replied in high rhetorical style: "I prefer to carry joyful rather than sad news to the most serene ears of my lord. But, when I see my monks wasting away from hunger and suffering from nakedness, how can I keep silent? . . . The storehouses and granaries have been emptied; in the purses there is nothing. What, therefore, am I, a sinner, doing here?"

Bobbio, indeed, had been stripped of its wealth before Gerbert arrived. The problem, he explained to Otto, was the "little books," or *libellarii*, a phenomenon unique to northern Italy. These were written contracts, valid in a court of law, by which the abbot gave the use of an estate, vineyard, or hayfield to a local lord in return for a percentage of the profits. Little books were intended to last twenty-nine years. Yet Gerbert's predecessors had renewed those that expired, and the lords had long considered them hereditary. As a result, when Gerbert arrived, much of Bobbio's land was in the hands of the Obertenghi, the descendants of "the illustrious Count Oberto," a family that supported Otto II when it pleased them. Otto expected Gerbert to redistribute this land to more loyal knights.

He tried, and the Obertenghi ignored him. Worse, they refused to pay him even what the contracts called for. Some sent payment to Gerbert's predecessor. Abbot Peter, from a noble Italian family, had become bishop of Pavia and Otto's chancellor, in charge of his treasury and correspondence. (He would soon advance to a still higher post: pope.) He apparently neither refused the gifts nor passed payment along to Gerbert. Peter had designated a monk named Petroald as his successor—before the emperor pulled rank and installed Gerbert as abbot—so some of the local lords paid Petroald, who was also from a well-known, noble Italian family (he may have been Peter's nephew). Like Peter, Petroald did not, at least at first, pass the money on to Gerbert.

Assuming Otto II would back him, Gerbert fought for Bobbio's rights with bluster and outrage underlain with threats. The castellan Boso felt himself entitled to a church and a hayfield. To him, Gerbert writes: "Let us avoid superfluous words and keep to facts. Neither for money nor for friendship will we give to you the sanctuary of God, nor will we consent if it has already been given to you by anyone else. Restore to Saint-Columban the hay which your followers took if you do not wish to test what we can do."

Writing to Bishop Peter of Pavia, Gerbert is equally strident: "You demand interviews yet you do not cease from thefts from our church; you, who ought to compel the complete restoration of what has been distributed, are yourself distributing our possessions to your knights as if they were your own. Steal, pillage, arouse the forces of Italy against us; you have found the opportune time. Our lord is occupied with the strife of war."

Gerbert's earnestness is admirable; it is also naïve. The thirty-year-old abbot is blunt, uncompromising, zealous, and impulsive, without any pretense of flattery; he is disrespectful, sarcastic, and, worst of all, clumsy.

He made grievous political errors. For instance, he rebuked Empress Adelaide, mother of Otto II, who wished him to give monastery lands as a benefice to her favorites: "I pray my lady to remember what she intimated to her servant—that she was about to ask in behalf of many persons more favors than could be granted." All the lands had already been assigned. Her favored knight, Grifo, had come too late. "If we give the whole away, what shall we keep? As far as it lies in our power, we shall do something for Grifo, but we will grant no benefice."

Even Emperor Otto felt the rough side of his tongue. Visiting the palace in Pavia while Otto was absent, Gerbert wrote to the emperor like teacher to student: "Why do the mouths and tails of the foxes here flatter my lord? Either let them depart from the palace, or let them present for judgment their satellites, who disregard the edicts of Caesar, who plot to kill his messengers, who compare even him to an ass. I keep silent about myself about whom they whisper in a new kind of talk. . . . The dispossessed have no sense of shame. O the times, O the customs. . . ."

Otto, understandably, was disappointed in Gerbert. He wanted shrewd administration, not lectures. He needed men-at-arms.

His hold on Italy was precarious. He was king through his mother, Adelaide, whose first husband had been one of several pretenders to the throne. When he had died, the seventeen-year-old Adelaide had been captured by "a man fierce and greedy, who would sell all justice for money," writes the medieval chronicler Widukind. When she refused to marry his son, he locked her in a tower. "Her flowing hair was pulled out," notes Odilo of Cluny, "her body frequently struck with blows from fists and feet." She escaped, hiding in a swamp for days until she was rescued by a fisherman.

Otto I, then king of Germany, took an interest in the beleaguered queen. "Hearing of her beauty and laudable reputation, [he] pretended that he was going to Rome," writes Thietmar of Merseburg. Passing through northern Italy, he secretly sent Adelaide a message. She agreed to his proposal, and they married in Pavia as soon as he had conquered it.

With Adelaide as his queen, Otto the Great soon ruled Italy from the Alps to Rome. He made a pact with the pope, who proclaimed him Holy Roman Emperor in 962. From Naples south, however, Italy was in the hands of Greeks and Byzantines (in principle, vassals of the Byzantine emperor), and, increasingly, Arabs—the writers of the time called them Saracens—expanding northward from their kingdom in Sicily.

Otto the Great had won several battles against the Greeks and Byzantines, pushing his borders south. Otto II meant to hold onto his father's conquests. In 981—shortly after installing Gerbert at Bobbio—he called up his German troops (few Italians joined his army). By June 982, he controlled the area south to Rossano, 300 miles from Rome, in the arch of Italy's boot. Leaving Empress Theophanu to hold the city, Otto pressed south again, this time facing Arabs.

He advanced to Stilo, well into the boot's toe, where he was stopped. The Saracens had seemed to flee before his armies. Chasing them, he landed in a trap. Says Thietmar, "Quite unexpectedly, they managed to gather themselves together and launch an attack on our forces, cutting them down with little resistance, alas."

The emperor, separated from his troops, fled on foot to the sea. A knight, recognizing him, gave up his horse, and the emperor swam it out to a passing Greek ship—which refused to take him on board. Returning to shore, he found the knight still there, "anxiously awaiting the fate of his beloved lord," writes Thietmar. In light of the future actions of Christian emperors, it is significant that the knight was named "the Jew Calonimus."

The emperor, continues Thietmar, "sorrowfully asked this man: 'What now will become of me?'" The Jewish knight urged him to swim the exhausted horse out to a second Greek ship. There, a Slavic knight recognized Otto. He hauled him aboard and hid him in the captain's cabin. The fate of the loyal Jewish knight is not told.

Soon, though, the Greek captain found the stowaway. After denying for some time that he was the emperor, Otto finally conceded, "Yes, it is I, reduced to this miserable state because of my sins." He would never again be king, he mourned. "I have just lost the best men of my empire and, tormented by this sorrow, can never again set foot in this land." But he had a plan: "Let us go to the city of Rossano where my wife waits my arrival. We will take her and all the treasure . . . and go to your emperor, my brother. As I hope, he will be a loyal friend to me in my time of need." The captain happily agreed, certain that the emperor of Constantinople—who was, as everyone knew, no friend to Otto— would richly reward him for these royal captives.

When they reached Rossano, the Slavic knight was sent to fetch Theophanu. She immediately understood the situation and made a plan. From the harbor, the Greeks saw her approaching with a train of sumpter mules presumably bearing the treasure Emperor Otto had mentioned. Theophanu sent a bishop and a few chosen knights on ahead, and the Greeks let them board. The bishop insisted Otto take off his bloody clothing before greeting his empress, and said they had brought Otto's robes of state. Otto got his drift. He stripped and, instead of reaching for the robes, "suddenly leaped into the water," Thietmar writes, "trusting in his own strength and skill at swimming," and so escaped the Greeks. Theophanu's knights drew their swords, and, "while the Greeks fled to the other side of the ship, our people followed the emperor in the boats which had brought them there, escaping without any injury." Theophanu promptly turned her

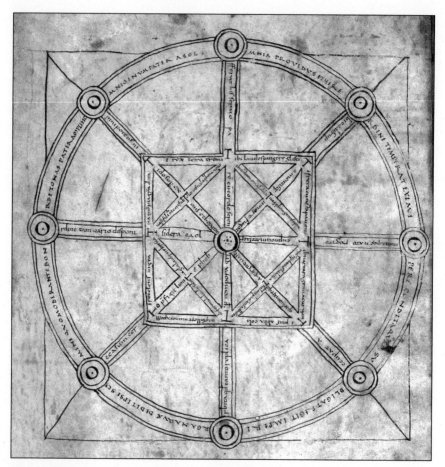

The puzzle-poem, or *Carmen Figuratum*, that Gerbert composed for Otto II in 983. It was presented as a booklet containing this image and thirty-two pages of explanations, but only the image itself has survived. It was copied into a music book made near Aurillac before 1079.

mule train around and took the emperor and their knights back to Rossano.

It was Otto's first defeat. He was astonished, his nerve destroyed. He returned to Rome and, for an entire year, did nothing but mope. The nobles of Germany and Italy regrouped. They sent a messenger to Otto, writes Thietmar, "with a letter that conveyed their humble desire to see him again. . . . The emperor agreed to this demand." They met at Verona in May 983, and Otto III—three years old—was declared king of Germany. He was sent north in the care of Archbishop Willigis of Mainz, who was charged with seeing the toddler crowned in Aachen.

Gerbert would have been among the vassals called to Verona, but he did not dare go. He feared being seen as a traitor. He had not sent the knights of Bobbio when the emperor called up his army—keeping them home to guard his hayfields instead. He felt, in some bitter way, responsible for Otto's defeat at the hands of the Saracens.

To mend their friendship, Gerbert began work on a great gift. In Mantua, in June, he met Otto and delivered his *Carmen Figuratum*, or "figurative poem," and a pipe organ. Writing to Archbishop Adalbero, he seems quite pleased on his own account, though he does not mention anything specific. "What I failed to accomplish at Mantua in regard to your affairs," he says, "I can explain to you better by words when present than by letters when absent." He closes with "Only absence from you disquiets our happiness day and night."

The *Carmen Figuratum*—a single page in which eight hundred red or black letters are arranged into the spokes and rim of a wheel, with two overlapping squares—was only identified as Gerbert's work in 1999. Ten years later, scholars are still unraveling the poem's many meanings, for Gerbert's artistry is not obvious.

Poems like this were complex word puzzles embedded in a picture. They had been court fashion in Emperor Constantine's day, and the library of Reims had a collection by Porphyrius, Constantine's chief poet, that Gerbert would have seen. Charlemagne had revived the art; his court mathematician Hraban Maur made poems in which a carpet of letters became suddenly readable when oriented to an overlaid cross or a series of Greek letters.

The standard way of presenting such puzzles was in a booklet. The picture would constitute the cover. Inside would be thirty-two pages explaining how to read the poem, and the various meanings to be derived. All this is missing for Gerbert's *Carmen*. Only the single cover page remains, copied to make the frontispiece of a *gradual*, or book of liturgical music. Because it includes music for the feast day of Saint Gerald the Good, the gradual is thought to come from Aurillac. It was made between 960 and 1079, but can't be dated more precisely.

This odd piece of art seems to have nothing to do with the music following it. Yet it is not pure accident that it was preserved—unlike the copy of Gerbert's abacus in the Giant Bible of Echternach. The wheel that holds the poem is drawn by the same hand that copied the rest of the music. And on the surface, at least, the poem seems to be about music: If you spin the wheel the right way up, the first word at the top, following a big O, is *Organa*, "organ." If you know how these puzzle-poems work, with layers and layers of acrostics and anagrams, you will soon realize that the lines of words intersect, always, at either a red "O" or a red "T." Eight O's circle the rim, with one at the hub. Twelve T's define the squares. Playing with the direction in which you read, you will come up with sixteen lines of good Latin beginning and ending with O, and sixteen lines beginning and ending with T.

Reorganized, the thirty-two lines of the wheel make eight stanzas in which the beginning and ending letters spell "OTTO." Read in order, these lines explicitly offer a pipe organ as a gift to the emperor: "This organ, which for my part I offer as a good omen, you . . . will make resound throughout the vast universe. . . . Your devoted abbot honors you."

Gerbert's meaning does not stop there: His *Carmen Figuratum* is a tour de force. To follow him further, you need to know that Hraban Maur often hid his signature in his poems, with the letters of his name making a picture or a symbolic shape. If you shuffle the thirty-two lines of Gerbert's wheel (preserving the "OTTO" motif, but not worrying if the sequence of lines make sense), you will find a complicated squiggle that reads *Gerberto Ottoni*, "from Gerbert to Otto." The shape is a Celtic knot used frequently in decorations on manuscripts and stonecarvings at Bobbio, which was founded by an Irish saint. In Celtic lore, the knot symbolizes the three faces of the goddess: virgin, mother, crone. In the manuscripts of Bobbio, it is used with a Christian meaning, alluding to the Trinity. To Gerbert, it was a symbol of fidelity, of a promise.

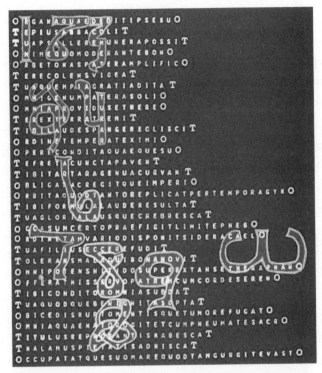

The Arabic numerals hidden inside Gerbert's *Carmen Figuratum*. The omega spells "and Theophanu," while the Celtic knot says, "from Gerbert to Otto." Another poem is created by the letters enclosing the numerals. It lists the names of the numerals and refers to the abacus and Gerbert's textbook on geometry.

In another squiggle nearby, you will find *ac Theophano*, "and Theophanu." This squiggle is shaped like the Greek letter omega. The omega holding Theophanu's name embraces the four lines that make up the name OTTO, as if the empress were embracing her emperor. Omega is, moreover, the symbol for eight hundred: A.D. 800 was the year Charlemagne became emperor, making the omega also, in Gerbert's mind, a symbol of empire. Not coincidentally, the poem has exactly eight hundred letters.

Given these two shapes, if you know how Gerbert's mind works, you might guess what to look for next—and you will find the nine Arabic numerals, in order, shaped just as they are on Gerbert's abacus board. The Latin words embedded in the numeric shapes make a grammatically sound poem in which Gerbert asks that the "guilty one" be forgiven. It ends with the clear statement, "Gift of Gerbert to Otto and Theophanu."

But that is not the end of the puzzle. If you look closely at the sequence of forty-six letters that make up these embedded words, you will find two more verses in Gerbert's style—in very good Latin—that seem to say, "Rumors have thrown my rarest things to the dogs of darkness. Ah! Could I have known you, in your mystery, from Odo?" Here, Gerbert is being purposely cryptic. This message was meant for Otto's eyes alone. The "dogs" are the "foxes" of his earlier letter to Otto warning about flattery: the men of the court. The reference to Odo could mean the king of France, anointed in 888, or perhaps "Oto," which he uses elsewhere in the poem to mean Otto I (he also refers to "Ottto," meaning Otto III).

The "rumors" verses are meant to be unclear: They hide an anagram. Cut up and put into a grid, and expanding the abbreviations, they produce twelve more lines of poetry. These verses contain delicate allusions to Otto and Gerbert's lives. Gerbert refers to the defeat in the south of Italy. He speaks sadly of his emperor, married to a Byzantine princess, who is yet attacked by Byzantines. He speaks of Theophanu as being like Venus, born of the sea. He also makes reference to the sciences of the quadrivium, mentioning the three genres of music and the orbits of the sun and stars. He refers directly to Arabic numerals, calling them (as al-Khwarizmi did) Indian. He names himself "practically the sole master" of this "Indian wisdom of numbers." He writes, "Gerbert will clearly instruct you so that you will understand these things having to do with the ten numerals," asking the king to pay attention to "the pronunciation of the Indian words: *igin, andras, ormis, arbas, guimas, calctis, tsenis, temenias, cerentis, sipos.*" These ten words, Gerbert explains in the poem, stand for the digits used in finger counting and the counters of the abacus board. "They are the names of my Geometry, august King Otto, but what is mine, in reality, is yours."

To those who can decipher it, Gerbert's *Carmen Figuratum* is a beautiful, profound poem, full of symbolism, mysteries, and cryptic messages.

Otto would have had no trouble working out its many meanings: In his presentation booklet, Gerbert had provided the key. To make the numerals, you must read the poem as a sequence of four letters "K." These stood for Constantine, Charlemagne, Caesar the Father, and Caesar the Son—Gerbert's usual way of designating the two Ottos—all spelled, as in Latin, with K. Both Ottos felt that they were heirs to Constantine and Charlemagne, destined to rejuvenate the Roman Empire. Encouraged by Gerbert, Otto III would later take this concept to extremes.

In this hidden poem, Gerbert says to Otto II, "The Empire spreads over all the cosmos." The emperor—the eight stanzas that each spell OTTO—is the guarantor not only of the poem's structure, but of the structure of the universe. And yet something is missing: number. Gerbert's point is symbolic. Without number, the cosmos becomes chaos. For this reason, when you reorganize the verses to reveal the shapes of the Arabic numerals, the lines about the devoted abbot's gift of an organ no longer make sense. If Otto concerns himself with such petty things as his servants' apologies and gifts, he will never rule the Roman Empire. To do so, Otto must learn to focus on number.

Gerbert is excusing the emperor's recent defeat in the south of Italy as a lack of knowledge. If Otto only understood numbers, Gerbert is arguing, he would be invincible. And he, Gerbert, "the master of number," can teach Otto what he needs to know.

But it was too late. Gerbert's bid to resume his place at court as the emperor's teacher and adviser failed. His poem, though happily received, had no effect. Six months later, in December 983, Otto suddenly caught a fever and died.

Hearing the news, the nobles surrounding Bobbio revolted. In fear of his life, Gerbert fled to Pavia, to a palace apartment owned by the monastery. From there he wrote, humbly now, to Empress Adelaide, begging for her protection. "Many, indeed, are my sins before God; but against my Lady, what, that I am driven from her service? . . . I thought I was practicing piety without avarice."

She ignored him.

He wrote to that old fox, Peter of Pavia—Gerbert's enemy had become Pope John XIV the previous July. "Whither shall I turn, father of the country?" Gerbert asks, in a small voice astonishingly unlike the tone he took earlier, when accusing Peter of stealing from his church. "If I call upon the Holy See, I am mocked and without opportunity to go to you." Just as Gerbert had gracelessly refused Peter's requests for an interview, the new pope denied Gerbert's. Gerbert suggested an intermediary: a mutual friend, the niece of Adalbero of Reims, Lady Imiza. "We love Lady Imiza because she loves you. Let us know through her, pray, either by messengers or by letters, whatever you wish us to do."

Imiza had married a duke and was one of the court ladies of Empress Theophanu. Gerbert wrote to her: "I consider myself fortunate in being accepted as a friend by such a remarkable woman as you. . . . Though Your Prudence does not need reminding, yet, because we feel that you are grieving and suffering severely from our misfortune, we wish the Lord Pope to be approached by messengers and letters, both yours and ours."

He wrote to the monk Petroald—whom Peter had preferred as abbot—and dumped Bobbio into his lap: "Do not let the uncertainty of the times disturb your great intelligence, brother," he said. "Chance upsets everything. Use our permission in giving and receiving, as becomes a monk and as you have known how to do. Do not neglect what we have agreed upon in order that we may have you more frequently in mind."

His gift for friendship had not completely deserted him in Italy. He and Petroald had come to respect each other, and he wrote soothingly to a monk named Rainard: "I urge and advise you to think and act as best you can according to your knowledge and ability. . . . Bewail the future ruin not so much of buildings as of souls; and do not despair of God's mercy." Five years later, he would ask Rainard to have copies made, "without confiding in anyone," of certain books in Bobbio's library.

And his knights seemed loyal. He wrote to Aurillac, "It is true, [they] are prepared to take up arms and to fortify a camp. But what hope is there without the ruler of this land, since we know so well the kind of fidelity, habits, and minds certain Italians have?"

Unsure where to turn, he asked Abbot Gerald if he could resume his studies with his former master—perhaps Raymond could meet him at Reims or Rome? He wrote to Miro Bonfill in Spain that he was ready to comply with his orders and suggested that Miro—who had unfortunately just died—also contact him at Reims or Rome.

But Rome was not really an option: The pope still refused to see him.

So Gerbert rode north to Reims, crossing the Alps again in January 984. Welcomed warmly by Archbishop Adalbero, he felt secure enough to subtly threaten the pope: "Deign to intimate to the holy bishops with what hope I may undergo the danger of approaching you. Otherwise, do not wonder if I attach myself to these groups where human but never divine law is the controlling factor."

For in Pavia—perhaps thanks to the intercession of Lady Imiza—Gerbert had made a secret arrangement with the Empress Theophanu. He would be her man in Reims: her advocate and spy.

Only twenty-three when Otto II died, Empress Theophanu was trapped between Otto's mother Adelaide, who had never liked "that Greek woman," and the German nobles, who held her little son hostage. When the three-year-old Otto III was consecrated king of Germany at Christmas, no one yet knew of his father's death. Writes Thietmar of Merseburg, "At the conclusion of this office, a messenger suddenly arrived with the sad news, bringing the joyous occasion to an end."

A regent would be needed until Otto III came of age. Theophanu saw herself as the most likely candidate: In the Byzantine Empire, an empress automatically ruled for her young son. The German nobles, at first, disagreed. Archbishop Willigis of Mainz, charged with the boy's upbringing, promptly turned his ward over to Henry the Quarreler, duke of Bavaria.

Henry seems an odd choice to trust with the life of Otto III: He had rebelled against Otto II seven years before and had been in prison for treason ever since. Yet many Germans wanted to separate their king-dom from Italy. They wanted a German king, not a Greek empress and a half-Greek boy. Learning of Otto II's death, the bishop of Utrecht, Henry's jailer, had immediately freed the duke—and urged him to be-come king. Egbert of Trier, the great friend of Adalbero of Reims, backed him as well, along with other important bishops. Willigis—though he gave up the boy—insisted that Henry could be *regent*, ruling until Otto III grew up, but not king.

The influential Notger of Liege also vacillated, and it was to him that Gerbert—master of rhetoric—wrote his first persuasive letter on Theophanu's behalf: "Are you watchful, O father of the country, for that onetime famed fidelity to the camp of Caesar, or do blind fortune and ignorance of the times oppress you?" Both "divine and human laws are being trampled underfoot," Gerbert warned. "Behold, openly de-serted is he to whom you have vowed your fidelity on his father's ac-count and to whom you ought to preserve it once vowed."

Gerbert then appealed to Notger's personal interests. He had certain knowledge, he wrote, that Henry the Quarreler and King Lothar of France planned to meet on the banks of the Rhine. There, Henry would give away the duchy of Lorraine—over which France and the empire had long contended—if Lothar would support his bid to be king. Notger's own archbishopric, Liege, was in Lorraine. His nobles would be dispossessed, their castles and estates bestowed on Lothar's Franks. Even church properties would change hands. Adalbero of Reims—born in Lorraine and brother to the count of Verdun—was against Henry's plot, Gerbert pointed out. Though a vassal of Lothar, Adalbero believed Lorraine rightly belonged to the Holy Roman Em-pire. He also knew a war between France and the empire would be dis-astrous for both.

Notger saw Gerbert's point. Joining with Adalbero, he convinced Lothar not to meet Henry or to accept his offer of Lorraine. They ap-pealed, in part, to the king's vanity. King Lothar, they insinuated, was more worthy to be regent than this duke. Both were related to Otto III to the same degree: Lothar's mother was sister to Otto I; Henry's father was brother to Otto I. Plus, Lothar was married to Emma, a daughter of Empress Adelaide by her first marriage. Why should he back a man of lesser rank whose claim was weaker than his own?

Nor was Lorraine a prize. Ruling it brought a risk to Lothar's dynasty. The current duke of Lower Lorraine was Lothar's brother Charles, and the two were not on good terms. Five years previously, Charles had spread the rumor that Lothar's queen was having an affair with Bishop Ascelin of Laon, the nephew of Adalbero of Reims, and Lothar had chased his brother from France. Otto II took Charles in and made him a duke. Provoked, Lothar attacked the imperial city of Aachen—where Theophanu was awaiting the birth of a child. Otto II swept his wife out of the city by night, and Lothar promptly sacked it. As a last insult, he had his men turn Charlemagne's great bronze eagle around to face France. With Theophanu safe, Otto II marched in revenge. From Aachen to Paris he pillaged the castles, but spared the churches.

Archbishop Adalbero had soothed the humiliated Lothar. Let Otto keep Charles and Lorraine, he argued. As a vassal of the emperor, Charles was no longer eligible to be king of France if anything (God forbid) should happen to Lothar. But if France took over Lorraine, Adalbero warned, Charles could threaten Lothar's throne—or that of his son, Louis, who had been crowned co-king at age twelve.

Now, with Otto II dead, Adalbero asked his king a question: Did France really want the warlike Henry on her border, when she could have the little child Otto under the regency of his gentle mother?

Three months later, Gerbert sent a letter to Lady Imiza: "Approach my Lady Theophanu in my name to inform her that the kings of the French are well disposed toward her son, and that she should attempt nothing but the destruction of Henry's tyrannical scheme, for he desires to make himself king under the pretext of guardianship."

At the same time, Adalbero set to work on his friend Egbert of Trier, also in Lorraine, using Gerbert to write the letters: "That your state is tottering through the cowardice of certain persons fills us not only with horror but also with shame. . . . Whither has sacred fidelity vanished? Have the benefits bestowed on you by the Ottos escaped from your memory? Bid your great intelligence return; reflect on their generosity, unless you wish to be an everlasting disgrace to your race."

Willigis of Mainz they also tried to turn from Henry's side: "With great constancy must we work, father, in order to maintain a plan of peace and leisure. What else does the disorder of the realms mean than the desolation of the churches? . . . Deprived of Caesar, we are the prey of the enemy. We thought that Caesar had survived in the son. O, who has abandoned us, who has taken this other light from us? It was proper that the lamb be entrusted to his mother, not to the wolf." Willigis soon joined their coalition, and brought many of his fellow Germans with him.

In June 984, with the king of France set against him and his support among the clergy having evaporated, Henry the Quarreler met Theophanu in Germany and surrendered little Otto III to her. Theophanu would reign as *Theophanius imperator augustus*, "Emperor Augustus," regent for her son, until her early death seven years later. Her mother-in-law, Empress Adelaide, would then take up the regency until Otto III came of age. For ten years the Holy Roman Empire would be ruled by a woman.

In July 984, Gerbert wrote to his agent at the palace looking for his reward: "Lorraine is witness that by my exhortations I have aroused as many persons as possible to aid him [Otto III], as you are aware." What plans did the empress have for him now? Should he remain in France "as a reserve soldier for the camp of Caesar"? Should he join Theophanu's court? Or should he prepare "for the journey which you and my lady know well enough about, as it was decided in the palace at Pavia"? That journey was back to Bobbio, where, with Theophanu's troops, he could regain control as abbot.

Six months later Gerbert was still in France, making himself useful at Reims and stewing over the empress's failure to answer. As he wrote to Raymond at Aurillac: "For these cares philosophy alone has been found the only remedy. From the study of it, indeed, we have very often received advantageous things; for instance, in these turbulent times, we have resisted the force of fortune violently raging not only against others but also against us."

Yet he was not satisfied. He had been an abbot—and a count.

Should he go to Spain, he wondered, or continue to wait for Theophanu's promised reward? To Abbot Gerald, he lamented: "Blind fortune, pressing down with its mists, enwraps the world, and I know not whether it will cast me down or direct me on, tending as I am now in this direction, now in that."

In the end, he stayed with Adalbero at Reims. While there, to keep himself busy, he resumed his teaching. As he wrote: "I offer to noble scholars the pleasing fruits of the liberal disciplines to feed upon." He never returned to Bobbio. For years he would mourn "the organs and the best part of my household paraphernalia" that he had left behind. He planned to fetch them "when peace has been made in the kingdoms," a time that would never arrive.

CHAPTER X

Treason and Excommunication

When Otto II died, the nobles of Bobbio were not the only Italians to rebel. Gerbert escaped with his life. The pope was not so lucky. As soon as Theophanu took her German army north, to reclaim her son and establish her regency, Pope John XIV, that old fox Peter of Pavia, was kidnapped. Peter had been Otto's chancellor. Though aristocratic, Italian, and qualified to be pope, he was perceived as Otto's creature. He was locked in the dungeon of the Castel Sant'Angelo, the fortress beside Saint Peter's in Rome. His captor was Pope Boniface VII. Boniface had been elected pope in 974 with the backing of the powerful Crescentian family (to make room for their favorite, they had strangled the sitting pope, Benedict VI). Evicted from office by Otto's army, Boniface robbed the Vatican treasury and fled with the money to Constantinople. Upon Otto's death, he returned and, with the Crescentians' help, reclaimed his seat. Peter of Pavia died in the dungeon in late 984.

Gerbert was horrified—and a tiny bit pleased. "The world shudders at the conduct of the Romans," Gerbert, then in Reims, wrote a Roman deacon with whom he shared books. "What sort of death did he suffer, that special friend of mine to whom I entrusted you?" Peter was hardly Gerbert's friend, and certainly not a "special" one.

Gerbert would soon learn not to gloat over an enemy's misfortunes. France was in turmoil. Adalbero, as archbishop of the chief city of France, Reims, was in the midst of it, and Gerbert was soon drawn into his intrigues. Again, Gerbert's forays into politics would bring him fame and power. Again, he would end up fleeing for his life—this time with his health ruined and under sentence of excommunication from the pope. In between, he would twice face being hanged for treason.

His fortunes are chronicled in his letters. Between January 984, when he returned to Reims, and February 996, when he left it in disgrace, Gerbert wrote, and kept, no fewer than 180 of them.

Some were the letters of a scientist and scholar. He asked for books. He discussed rhetoric and organ-playing with the monks of Aurillac, wrote to Spain to learn more about mathematics, explained the abacus to Remi of Trier, and discussed climate circles and the making of celestial spheres.

Others were the letters of a friend: He promoted his friend Constantine as a music teacher and attempted to pull strings to get him elected abbot of Fleury. Alas, their mutual enemy Abbo was chosen instead.

His longest letters were those he wrote to justify himself—particularly, his defiance of the pope when the archbishopric of Reims became a pawn in the battle for France.

But most he wrote under commission: The master of rhetoric composed letters in the name of Archbishop Adalbero, Duke Charles of Lorraine (the pretender to the French throne), Count Godfrey of Verdun (Archbishop Adalbero's brother, then imprisoned by King Lothar of France), Queen Emma (King Lothar's wife and the daughter of Empress Adelaide), and Duke Hugh Capet (who would become king through Adalbero's machinations). He wrote to Empress Theophanu, Empress Adelaide, King Lothar, Charles again, and the Byzantine emperors; to dukes and duchesses, counts and countesses; to archbishops, bishops, abbots, and monks throughout France, Germany, Italy, and Spain.

Occasionally he was messenger as well: Having written a letter, he took it to its intended recipient and waited, pen in hand, to help shape the reply. From Bishop Dietrich of Metz to Duke Charles of Lorraine, he wrote: "You fickle deserter, keeping faith neither in this direction nor in that, the blind love of ruling drove your weak-minded self to neglect a pledge, given under oath before the altar of Saint John. . . .

Have you ever had any scruples? Swell up, grow stout, wax fat, you who, not following the footsteps of your fathers, have wholly forsaken God your Maker."

From Charles to Dietrich, Gerbert wrote: "It has befitted my dignity, indeed, to cover up your curses and not to give any weight to what the caprice of a tyrant rather than the judgment of a priest proffered. But, lest silence would imply to your conspirators the making of a confession, I shall touch briefly upon the chief particulars of your crimes, saying the least about the greatest. Grown stout, fat, and huge, as you rave that I have, by this pressure of my weight I will deflate you, who are blown up with arrogance like an empty bag."

Gerbert learned tact. Under Archbishop Adalbero's tutelage, the clumsy courtier from Bobbio grew into a slick and crafty flatterer. He also became a spy. He frequently slipped word to Theophanu—his sworn overlord, after all—of the weaknesses of the French. Through intermediaries, he advised her to ally with Hugh Capet, not the weak and vacillating King Lothar. "Lothar is king of France in name only," he wrote, "Hugh not in name, it is true, but in deed and fact."

With such letters circulating, it's no surprise that Lothar and his son, Louis V, accused Gerbert and Adalbero of treason. But alienating the archbishop of Reims was a mistake. It would cost the two kings of France their lives, as Adalbero—with Gerbert's help and the secret intervention of Theophanu—put Hugh Capet on the throne. In 987, the line of Charlemagne came to an end.

Hugh Capet was born in 940, making him about ten years older than Gerbert. Lothar was born a year later. Lothar and Hugh had long been rivals. Hugh was descended from four French kings; Lothar was descended from Charlemagne. All else being equal, the French preferred a Carolingian king. But when the heir of Charlemagne was too young or weak, they had crowned Hugh's ancestors, beginning with King Odo in 888. King Lothar came to the throne as a thirteen-year-old, in 954, only because Hugh's father declined to challenge him.

A tiny ivory carving made for the cover of a book shows Otto II and Theophanu being blessed by Christ. Their names, written in a mix of Latin and Greek letters, are given as "Emperor Otto" and "Emperor"— not Empress—"Theophanu," the title Theophanu assumed as regent for her son, Otto III.

Hugh Capet succeeded his father as Duke of France in 956, when he was sixteen. As duke, he controlled more land than the king and fielded more knights. The duke of France was considered the king's right arm—and so Hugh was indignant when Lothar secretly made peace with Emperor Otto II in 978, after Lothar had sacked Aachen and Otto had retaliated by marching on Paris. Excluded from the treaty, Hugh decided to make one of his own and rode to Rome.

As Richer of Saint-Remy tells it, Otto kissed Hugh and, setting aside their differences, treated him as his sweetest friend—except that he spoke in Latin, which Hugh did not understand. Bishop Arnulf of Orleans, Hugh's confidant, translated. Their pact agreed, Otto left the room, then popped back in. He had left his sword lying on a bench. "Would you hand it to me?" he asked Hugh. The bishop intervened— "Oh, allow me!"—and grabbed it. Otto grinned, thanked him, and left. The bishop's quick thinking, Richer said, saved Hugh from inadvertently declaring himself the emperor's vassal, in front of everyone, by carrying Otto's sword to him. But had it? Simply by riding to Rome and presenting himself as a supplicant, Hugh had shown himself willing to do Otto's bidding.

Lothar and his queen certainly saw Hugh as a turncoat, conspiring with the emperor against them. Queen Emma, says Richer, wrote to her mother, Empress Adelaide, asking her to seize Hugh on his way home from Rome. "And so that this man of bad faith does not escape you by his ruses, I have taken care to note down for you the essential characteristics of his physique." Richer does not quote more of Emma's letter, or we would know what Hugh looked like. He only paraphrases: "She continued by indicating the particularities of his eyes, his ears, and his lips, even his teeth, his nose, and the other parts of his body, not forgetting the way he talked." Hugh was wilier than suspected. Disguised as a stablehand, he made his way north caring for his own pack horses. Once safe in Paris, he pretended nothing had happened.

But Lothar knew he could no longer depend upon the duke of France. Trying to bind other knights closer to him, in 979 Lothar married his son Louis (who had just been crowned co-king), to Azalais, sister of the powerful count of Anjou. Twice widowed, Azalais herself controlled much of Aquitaine, an enormous duchy then covering most of southern France, and the duchy of Burgundy, which bordered the kingdom of Burgundy on the Italian frontier. The marriage failed. The bride was "an old woman" of thirty; the groom a boy of fourteen. "Of conjugal love between them," says Richer, "there was none. They refused to share a common bed and, when retiring, would not even sleep under the same roof. When they had to meet, they chose a spot out of doors. Their conversations were limited to a few brief words. This lasted for nearly two years."

Nor did Louis impress Azalais's knights. At fourteen, his father had been king; Louis, Richer laments, remained a wastrel. "The situation became deplorable. Louis shamed and discredited himself by his inability to govern." He lived in a "miserable manner, destitute and in distress, lacking both personal wealth and military force," until his father finally called him home. Azalais took advantage of her husband's absence to marry herself to Count William of Arles, in the independent kingdom of Burgundy. Louis earned nothing but the epithet "Louis Do-Nothing." Not only the duchies of Aquitaine and Burgundy, but the county of Anjou slipped from King Lothar's control. Rather than strengthening his hold on the throne of France, his son's failed marriage made it weaker.

Forced to abandon his plans in the south, Lothar looked east, again setting his sights on the rich duchy of Lorraine. With the death of Otto II in 983, Lothar saw his opening. He entertained the idea of a treaty with the pretender to the German throne, Henry the Quarreler. As soon as Henry submitted to Empress Theophanu in 984, Lothar instead declared war. He called up his troops—Hugh Capet, among others, did not respond. He crossed the border into the empire and sacked the city of Verdun.

Archbishop Adalbero's position was delicate. His brother, Godfrey, was count of Verdun and a vassal of Theophanu. When the city fell, Godfrey and one of his sons were captured by Lothar's troops. Godfrey's wife and two other sons escaped to lead the resistance. But Adalbero, as the archbishop of the chief city of France, was Lothar's premier counselor. He was also Lothar's chancellor, in charge of his correspondence and his treasury.

Lothar decided—unwisely—to test his counselor's loyalty. He sent Adalbero to Verdun to hold the city for France while he attacked Liege. Adalbero spent most of his time there trying to have his brother's third son appointed as Verdun's bishop—he had been nominated by Empress Theophanu. Adalbero's friend, Egbert of Trier, refused to risk King Lothar's anger by consecrating him. "Render unto Caesar the things that are Caesar's," Adalbero self-righteously admonished Egbert (through Gerbert's pen), "and unto God the things that are God's."

It was a smokescreen. While Adalbero made a fuss over who should be bishop of Verdun, Gerbert had secretly gained access to Count Godfrey, Lothar's prisoner, and was running messages for him to Lothar's enemies. These letters are clear admissions of treason—so clear, it is surprising Gerbert kept copies of them. If any had fallen into King Lothar's hands, both Adalbero and Gerbert would have been hanged. Their support for Theophanu and her vassal, Count Godfrey of Verdun, are clear. As is their opinion of King Lothar and his French troops: They are "the enemy."

To Godfrey's wife, for example, Gerbert wrote: "You with your sons preserve an unstained fidelity to Lady Theophanu. . . . Make no agreement with the enemy, the French; repulse the kings of the French; so keep and so defend all the forts that the latter may not think you have abandoned any part of your followers in these forts, neither because of the hope of freeing your husband, indeed, nor because of the fear of his death." Referring to Lothar's attacks on Verdun and Liege, Gerbert wrote to one of Godfrey's allies, "By no means should you believe Adalbero, Archbishop of Reims, and most loyal to you, to be an accomplice of these deeds, for he himself is oppressed by great tyranny." Adalbero is loyal, that is, to Lorraine, not to France; to Theophanu, not to the tyrant, Lothar.

In April 985, Gerbert told Theophanu that he and Adalbero had been found out. Lothar accused them of treason and threatened them with death. Gerbert wrote to the empress, "Where and when we can go to your presence, if any road through the enemy shall be open, indicate to us more definitely. . . . Matters have reached this point that it is no longer a question of his [Adalbero's] expulsion, which would be an endurable evil, but they are contending about his life and blood. The same is true of myself."

There is no record of Theophanu's reply, but it appears likely she followed Gerbert's earlier advice and made an offer to Hugh Capet. In May, when King Lothar convened his nobles in Reims to try Adalbero and Gerbert for high treason, Hugh arrived with six hundred knights— far more than the king could command. "This report suddenly dissolved and scattered the meeting of the French," Gerbert crowed. With Hugh Capet shielding the traitors, Lothar could not proceed. The charges (though true) were dropped. To Adalbero, in hiding, Gerbert wrote, "Hugh's friendship ought to be actively sought after, and every effort should be exerted lest we fail to make the most of this friendship which has been well begun."

Adalbero returned to Reims as if nothing had happened, and Gerbert secretly began writing to Hugh's sister Beatrice, duchess of Upper Lorraine. She was his new channel to Theophanu. First he instructs Beatrice in the art of intrigue: "Secret information certainly ought not to be entrusted to many persons; but not without cause do we assume that letters written to us in different handwritings have been handled by different persons." Then the news for the empress: "A plot either has been formed or is being formed against the son of Caesar and against you. . . . Through the adroitness of certain persons Duke Hugh was finally reconciled with the king and queen on June 18 in order to create the impression that such a great man's name is promoting the plot—a very unlikely thing, and at this time we think he will not do so."

The winter passed and, as far as we know, the plot against Theophanu did not take shape. A counterplot, however, may have succeeded. In early spring, King Lothar fell ill, with fever and vomiting, cramps and nosebleeds. He soon died. Some said he was poisoned, though no one dared say by whom. Gerbert, at Adalbero's direction, organized a magnificent funeral. The corpse was dressed in silk and covered with a purple cloak ornamented with precious stones and gold embroidery. Laid on a regal bed, it was borne on the shoulders of noblemen, with bishops and monks preceding it, chanting psalms and carrying the royal crown among crosses and holy books, a long cortege of woeful knights trailing behind.

Lothar's son, the nineteen-year-old Louis Do-Nothing, became king of France as Louis V. His mother, Queen Emma, used Gerbert as her secretary to ask her mother, Empress Adelaide, for aid. Deals got underway to ransom Count Godfrey of Verdun, to consecrate his son as bishop of that city, and to effect a lasting peace between King Louis V and Emperor Otto III (through their mothers). Suddenly Louis—influenced by his uncle, Charles of Lorraine—broke with his mother, accusing her again of adultery with Bishop Ascelin of Laon. His real target, however, was the bishop's uncle—Adalbero of Reims—and his accomplice, Gerbert.

"My sorrow has been increased, O my lady," Emma told her mother, through Gerbert's pen. "When I lost my husband there was hope for me in my son. He has become my enemy. My once dearest friends . . . have fabricated the wickedest things against the bishop of Laon, to my disgrace and that of my whole family."

Bishop Ascelin of Laon sought refuge with his uncle at Reims. Young King Louis proceeded to attack the city, destroying the bishop's palace. "In what full measure the wrath and fury of the king have burst forth against us is evidenced by his sudden and unexpected attack," Gerbert wrote in alarm to Theophanu. "We beg you to bring definite aid to us, therefore, at this uncertain time, permitting no false hope to delude us who never have hesitated in maintaining our fidelity to you. At a meeting of the French, planned for March 27, we are to be accused of the crime of treason." Louis did not intend to let the traitors slip away this time. He apparently knew more—and was less of a "do-nothing"—than Gerbert thought.

Again, Theophanu does not seem to have replied. Yet the coup that she and Adalbero had long plotted, as Gerbert's letters show, was about to succeed. Mysteriously, the meeting of the French nobility was postponed until May 18. Before the actual trial began, Louis went out hunting. According to Richer of Saint-Remy, whose history of France is dedicated to the archbishop of Reims, Louis fell off his horse and damaged his liver. With blood flowing from his nose and mouth and wracked with fever, he died. He was twenty years old. Other historians of the time claimed, again, that the king of France had been poisoned.

Hugh Capet took control of the assembly and dismissed the charge of treason against Adalbero and, by extension, Gerbert. "Since there is no one here to accuse him, we must find in favor of the archbishop," he reasoned, "for he is a nobleman of great wisdom." Adalbero then suggested the council reconvene in a week's time to elect a new king.

Charles of Lorraine—King Lothar's brother, King Louis's uncle, the last heir of Charlemagne—was the obvious choice. Adalbero nevertheless publicly rejected him. He was the vassal of a foreign ruler (Empress Theophanu). He had married a woman of low rank. He was "untrustworthy and indolent," and he had surrounded himself with perjurers and evil men—"nor do you want to part from them," Richer reports the archbishop saying. To the assembled nobles, this last objection was key. As king, Charles would want to reward his followers with lands and castles—but he owned none in France. He would have to take them from his brother Lothar's faithful noblemen. Many of those listening would have been dispossessed.

Concluded Adalbero, "The throne is not acquired by hereditary right, and we must elevate to it a man distinguished not only by the nobility of his birth, but also by the wisdom of his spirit, a man for whom honor is a shield and generosity a rampart." That man was Hugh Capet. Paris, seat of his duchy, would become the chief city of France, and the Capetian dynasty would rule for the next four hundred years.

Adalbero was restored to power as the king's first counselor, and Hugh began building him a new palace at Reims. Verdun was returned to the empress. Godfrey was released; his son was confirmed as bishop. Gerbert was hired to tutor Hugh's son Robert so that he could not be humiliated by any emperor, as his father once was, for his lack of Latin. As Hugh's secretary, Gerbert wrote to the Byzantine emperors seeking for Robert a royal bride. And he wrote to Count Borrell of Barcelona promising aid against the Saracens (for a price).

But before Hugh could ride to the rescue of Spain, he had to deal with Charles. As the last heir of Charlemagne, the last Carolingian, Charles had many friends. Instead of dating their records for 987, "In the first year of the reign of Hugh Capet," churches in the Limousin, Quercy, Poitou, Velay, and various other parts of France wrote, "Waiting for a king." The citizens of Laon, in the center of the Carolingian lands, opened their gates one night at dusk to let in the soldiers Charles had hidden in their vineyards. Bishop Ascelin and Queen Emma were captured. Charles fortified the walls and towers, dug ditches, built catapults, and set a guard of five hundred men furnished with crossbows. What began as a coup had turned into a civil war, a battle in which every French noble had to choose between the Capetian and the Carolingian.

Hugh Capet called up an army and besieged Laon. He offered to negotiate and sent in Gerbert, but Charles, "contemptuous of emissaries," refused to give up the town or ransom any of his hostages.

Gerbert returned to the king, who soon found other uses for his skills. Laon was on a hill and impregnable. One August day, "after noon, while the king's soldiers were deep in wine and sleep," Gerbert wrote, "the townspeople with their whole strength made a sally; and while our men were resisting and repelling them, these very ragamuffins burned the camps. This fire consumed all the siege apparatus." The siege weapons were rebuilt. Richer describes one in detail, and implies Gerbert designed it. Gerbert himself wrote, "The labor of the siege against Charles has exhausted me and violent fevers have been harassing me," while in Adalbero's name he confidently tells King Hugh, "Look for us with all the troops to break through the fortress and uproot the mountain from its very foundations, if this is your desire."

They failed. The autumn nights grew cold and Hugh lifted the siege until spring—at which point Bishop Ascelin bravely slid down a rope and escaped Charles's clutches.

Ascelin conferred with Adalbero. Hugh and Emma, through Gerbert's pen, consulted Theophanu. It seemed the battle for the throne of France would be a short one. With Charles locked in Laon, and Theophanu blocking any aid he might have expected from his duchy of Lower Lorraine, the Capetians had nearly won. At Reims, Gerbert finally had some spare time to work on the celestial sphere for Remi of Trier and, perhaps, to confer with Constantine about the astrolabe. Then suddenly his world was plunged "into primordial chaos": Archbishop Adalbero of Reims fell ill and died in January 989.

Gerbert wrote of crisis, confusion, and "great disturbance." For twenty years, Adalbero had been in charge of the cathedral of Reims and its related school and monasteries. He was the leading churchman of

France: Where Reims went, the bishops and abbots of France followed. As the king's chancellor, he had been in charge of the royal correspondence and treasury. As the king's premier counselor, he had shaped policy toward the empire and the papacy—lately, in spite of the actual king's wishes, but always, he would argue, in the best interests of France. Moreover, he was Gerbert's dear friend and mentor. "I was so suddenly deprived of him that I was terrified to survive, since, indeed, we were of one heart and soul," he told Raymond at Aurillac. "Heavy cares so weighed me down that I have almost forgotten all my studies," he confided to a monk named Adam.

He had another reason to be afraid: He was expected to take Adalbero's place. "I keep silent about myself for whom a thousand deaths were planned," he wrote to Remi, "both because Father Adalbero with the assent of the whole clergy, of all the bishops, and of certain knights had designated me as his successor; and because the opposition maintained that I was the author of everything that displeased them"—principally the election of Hugh Capet. "With pointing fingers they singled me out for the ill will of Charles, then as now harassing our land, as the one who deposes and consecrates," the kingmaker.

Without Adalbero, Gerbert was out of his depth. He had too little nobility of blood—no family alliances, no counts to call on, no network of henchmen—and too much nobility of soul. He had become an adroit spy, true. But he could justify that to himself: He had never sworn an oath of allegiance to the kings of France, as he had to Theophanu. In his fight for the archbishopric of Reims, however, his high-mindedness was a drawback. He was too honest to swear false oaths, expose his own sanctuary to plunderers, render pretend excommunications, or bribe the pope—as his rival did. He didn't have a chance. Yet it was his duty, he felt, to try.

Adalbero's seat, as archbishop of Reims, the leading churchman of France, had to be filled. Gerbert's opponent was Arnoul, the twenty-four-year-old bastard son of King Lothar, and thus Charles's nephew. Educated at Reims by Gerbert, Arnoul was a cleric at Laon, now under Charles's rule.

Hugh pondered his choice. He discussed it with Gerbert. "King Hugh . . . offer[s] much, but we have received nothing definite thus far," Gerbert wrote to Willigis of Mainz, now Theophanu's chief adviser, who had implied that a post might be found for him in the empire. At the same time, he called in what favors he was owed. A son of Count Godfrey of Verdun (through Gerbert's pen) sent Hugh a warning about Arnoul: "Do not think of putting in charge in that place a deceitful, ignorant, good-for-nothing who is unfaithful to you. . . . You should not wish to entrust your safety to the advice of those who have decided to advise nothing except with the assent of your enemies."

But by appointing a Carolingian, Hugh thought, he could end the civil war. Besides, Arnoul was the son of a king, Gerbert of a peasant. In April, Arnoul swore an oath of fidelity to King Hugh and became archbishop of Reims. As the archbishop's official secretary, Gerbert had to write the proclamation.

Distraught, he asked the archbishop of Mainz about the once-promised post in the empire: "Pray remind my Lady Theophanu of the fidelity that I have always maintained toward herself and her son. Do not allow me to become the prize of her enemies whom I reduced to disgrace and scorn on her behalf whenever I was able." Theophanu did not answer.

Six months later, Arnoul broke his oath to King Hugh. He called a council of nobles at Reims and, in the dark of night, instructed a priest named Augier to open the gates. Charles's army streamed in. They sacked the town and pillaged the cathedral, took all the nobles hostage—including Arnoul—and locked them up in Laon. Gerbert they left to look after Reims, implying that he was their accomplice. Arnoul, pretending innocence, excommunicated the "authors of the robberies at Reims" (again, Gerbert had to write the declaration): "May the eyes of you who coveted be dimmed; may the hand which looted wither; . . . may you dread and tremble at the appearance of an enemy . . . until you disappear by wasting away." Was he excommunicating himself? His crony Augier? Charles or the soldiers?

Then Arnoul swore a new oath to uphold Charles's right to the kingship. Clearly, Hugh had chosen the wrong archbishop for Reims.

Gerbert despaired. All that he and his beloved Adalbero had accomplished was in danger of being undone. Reims was in the hands of the enemy. "We have entered upon the restless sea," Gerbert wrote to an unknown friend. "We are shipwrecked, and we groan. Never do safe shores, never does a haven appear." One summer night, he slipped out of Reims and sought refuge with King Hugh. "Not for the love of Charles nor of Arnoul will I suffer longer to be made an instrument of the devil by proclaiming falsehoods contrary to truth," he wrote.

O nce again, the resourceful Bishop Ascelin of Laon turned the course of the war. Pretending to be dissatisfied with King Hugh, Ascelin weaseled his way into Arnoul's good graces. He was allowed to return to Laon to confer with his monks and knights and sort out their petty problems. He held a feast there for Arnoul and Charles, swearing on saints' relics that he meant no betrayal. As Ascelin reached for his wine, Charles said again, "Do not touch that cup if, like Judas, you mean to betray me." Ascelin took up the cup and drank.

Late that night, Ascelin snuck into Charles's and Arnoul's rooms. He took their weapons, then called in his knights. The last two Carolingians were captured and hauled off, naked, to a tower prison; the town gates opened to Hugh's men. Charles would die in prison two years later, aged forty, possibly poisoned like his kinsmen, Kings Lothar and Louis. The civil war between Hugh and Charles had come to an end.

But what was to be done with Archbishop Arnoul?

Even before Ascelin's trick, King Hugh had tried to consult the pope. "Aroused by new and unusual events, we have ordered that your advice must be most eagerly and carefully sought," Gerbert wrote for him. "Take under consideration what has been done, and write back in reply what ought to be done in order that respect for sacred laws may be revived and the royal power not be nullified." Gerbert also sent a letter in the name of the bishops of Reims province, alerting the pope to "the new and unprecedented crime of Arnoul, archbishop of Reims."

Arnoul's supporters, however, had reached the pope first—and brought him the gift of a beautiful white horse. The king's messengers were left waiting outside the Lateran palace until they gave up in disgust. The two letters were never acknowledged.

Eighteen months later, without the pope's knowledge or approval, Archbishop Arnoul was dragged before a council of French bishops at the monastery of Saint-Basle outside Reims. He was defrocked and forced to apologize to King Hugh and his son, Robert. Gerbert was then named archbishop of Reims in his stead.

Gerbert should have been ecstatic at his victory. He was not. He did not want to be archbishop—he wanted Adalbero back. The post was a duty laid upon him by his dear mentor, not a joy. It took four years for him to even share news of his great honor with his friends at Aurillac. He was "distracted to the utmost by the preoccupations of important business," he writes, explaining his failure to write. But Gerbert accomplished little in those years—or over the rest of his term. Unlike Adalbero, he did not redo the cathedral, reform the monasteries, or rejuvenate the cathedral school. Instead, he spent all his energy defending himself. His election, he wrote to his friends, "aroused races and peoples to hate me." His opponents cited church law to try to reinstate Arnoul. "More tolerable is the clash of arms than the debates of laws," Gerbert writes. "Though by oratorical ability and a wordy explanation of the laws I have satisfied my rivals as far as it concerns me, thus far they have not yet abandoned their hatred."

Nor would they. For seven years—even after Gerbert had fled Reims—they fought his appointment. The position always would be a burden to him, a responsibility. He must have wished, more than once, that he was still just a schoolmaster, his only duty to enlarge minds.

The sides in the conflict, however, had been chosen long ago: They were the friends and enemies Gerbert had made as a scientist and scholar. Leading the fight against him was his old enemy, the loquacious and exasperating former schoolmaster who considered himself a "doctor of the abacus" and master of (at least) five of the seven liberal arts, the fierce reformer and protector of monks' rights, Abbot Abbo of Fleury. On the other side was Bishop Arnulf of Orleans, the confidant of Hugh Capet who had so quick-thinkingly snatched up Emperor Otto's sword in Rome, the friend and patron of Gerbert's beloved Constantine.

Arnulf and Abbo were themselves fierce enemies. Arnulf, as bishop of Orleans, was technically Fleury's overlord; Abbo refused to acknowledge it. Recently they had come to blows over control of a vineyard. To defend his monastery's rights to the grapes, Abbo had raised a "clamor," a specific and highly public form of request for saintly aid. First, he and his monks covered the floor of their church with sackcloth. They doused the candles and stripped the altar of crucifix and reliquaries, setting them on the floor in disgrace. Then the monks threw themselves face down while a priest called out: "Rise up in our aid, Lord Jesus! Comfort us and help us overcome our attackers, and crush the pride of those who afflict us and your monastery." The relics thus roused, Abbot Abbo marched them out to the disputed vineyard, where Bishop Arnulf had posted men to prevent the monks from harvesting the grapes. Confronted with superior holiness—or at least an effective spectacle—the bishop called off his men.

It wasn't the end of the dispute. Later, on his way to Tours to celebrate the feast of Saint Martin, Abbo was waylaid by the bishop's men. Swords were drawn. Some of Abbo's retinue were killed.

In 991, at the monastery of Saint-Basle outside of Reims, Bishop Arnulf and Abbot Abbo faced off again, this time over the archbishopric of Reims. The bishop accused Arnoul of treason. Abbo defended him.

In a spectacular conflict of interest, Gerbert served as the official secretary during the dispute. We have no copy of his original notes, only a revision he called *The Acts of Saint-Basle*, which he sent to the pope's legate in 995. He described the council again at the end of that year in a long letter to an adviser of Otto III.

Richer of Saint-Remy (who was not present) also told the story in his *History*, writing between 996 and 998. According to Richer, the priest Augier confessed that he had unlocked the gates of Reims on Arnoul's orders. Arnoul himself was asked "a great many questions; some of which he answered, and some refused," wrote Richer. "Finally, defeated, he succumbed to the logic of the argument and publicly confessed himself guilty and unworthy of the archbishopric."

King Hugh and his son Robert then joined the council. "'Prostrate yourself before your lords and before your kings, whom you have unforgiveably offended,'" Arnoul was commanded, "'and confessing your own guilt beseech them for your life.' And prostrating himself in the form of a cross and crying aloud, he so supplicated for his life and limbs that he reduced the entire synod to tears." Out of pity, Hugh and Robert spared him.

According to Gerbert, Arnoul was more wicked than pitiful: He "engages in arson, sedition, betrayals, disgraceful acts, captures, and thefts from his own men, while he plots his kings' destruction, and betrays his land to the enemy." He "confessed to those crimes," Gerbert points out, and is therefore "dead forever as a priest." Abbo of Fleury's efforts to have Arnoul reinstated as archbishop (and Gerbert removed), Gerbert claims, are due only to "consuming envy and blind cupidity."

But the biggest hole in Richer's story is his failure to mention the debate about the power of the pope—and Gerbert would have helped himself immensely if he, too, had omitted it from his account of the event.

Abbot Abbo, defending Arnoul with numerous citations of precedents and laws, did not contest his betrayal of Hugh Capet. But treason, Abbo said, was no grounds for defrocking an archbishop. (Five years earlier, Adalbero and Gerbert might have happily agreed with him.) Nor did a council of bishops have the right to do so *for any reason*. Only the pope could make or unmake an archbishop.

To Bishop Arnulf—and King Hugh—that idea was absurd. The pope! Who refuses even to answer the king's letter? The bishop responded with a speech so vehemently antipapal that it was once thought to be a forgery inserted by Protestants in the sixteenth century. "Deplorable Rome! . . . What spectacles have we not witnessed in our days!" he began, and regaled the council with papal deeds of debauchery, treason, violence, and murder—including the very recent kidnapping of Peter of Pavia by a rival pope. "Can bishops," concluded Arnulf, "legally submit to such monsters swollen with ignominy, lacking all sciences, both human and divine?" Just because a man sits on a throne, "resplendent in his purple and gold," should we listen to him? "If he is lacking in charity, if he is not filled with and supported by science, he is the Antichrist sitting in the temple of God. . . . If he is neither supported by charity nor held upright by philosophy, it is a statue, an idol in the temple of God. To ask it anything is to consult marble." To call the pope the Antichrist—or a marble statue—was rather extreme, and Gerbert (if not Arnulf) would come to regret it. But on the whole, Arnulf's accusations were true.

Tenth-century popes were not the powerful religious leaders of today. They were political pawns. Many were not even churchmen. For much of the century the papacy was influenced by the mercurial Roman noblewoman Marozia. She was mistress of Pope Sergius III (904–911), murderer of John X (914–928), and mother of John XI (931–935). Her grandson, John XII (955–963), was both pope and Prince of Rome until he double-crossed Otto I, whom he had just crowned emperor. At a synod in Saint Peter's, John XII was accused of sacrilege, simony, perjury, murder, adultery, and incest, and then deposed. He excommunicated the members of the synod, and when he caught three of them, he flogged one, cut off another's right hand, and the third's nose and ears. Otto I marched on Rome, but before he arrived John was "stricken by paralysis in the act of adultery" and died.

Otto's appointee, Leo, wasn't even a priest. The Romans chose Benedict, a deacon, who was well qualified. Pope Benedict was "attacked by Leo, aided by the emperor," Arnulf claimed. "Besieged, made prisoner, and deposed, [Benedict] was sent in exile to Germany," and Otto appointed John XIII, a bishop and, incidentally, Marozia's nephew. Pope John XIII was captured by a rival faction, but escaped.

The emperor hanged the conspirators, and John XIII went on to have a successful papacy. Bishop Arnulf found nothing ill to say of John XIII (at least in the version Gerbert recorded); this was the pope whom the young Gerbert, fresh from Spain, had impressed with his mastery of *mathesis*.

His successor, chosen by Otto II, was strangled by supporters of his rival, Boniface VII. This was the antipope who fled (first robbing the Vatican treasury) when Otto invaded the city, returned when Otto died, and threw Peter of Pavia into the Castel Sant'Angelo. When Boniface VII himself died a year later, his body was dragged through the streets of Rome by a mob.

The nobles of Rome, led by the Crescentians, replaced the antipope with a Roman nobleman, who became Pope John XV. By the time of Archbishop Arnoul's treason, John XV had reigned six years. He would last another five by carefully balancing the desires of Crescentius of the Marble Horse, Prince of Rome, with those of the empresses Theophanu and Adelaide.

Abbot Abbo made sure John XV heard Bishop Arnulf's antipapal diatribe.

Incensed, the pope dispatched his legate, Leo, to fling the "Antichrist" and "marble statue" insults back in the French bishops' faces. "You are the Antichrists, who say that the Apostolic Church is governed by an inert statue, by an idol resembling those of the pagans. Is there a single Christian who can listen in cold blood to such blasphemy? What! Because the vicars of the blessed Peter and his disciples study other masters than Plato, Virgil, Terence, and the rest of that troop of philosophers . . . you conclude that they are not worthy of being promoted to the ranks of doormen because they ignore the poets?"

Over the next five years, six more church councils met—in France, in Italy, and in Germany—to debate whether Arnoul or Gerbert was the true archbishop of Reims. The struggle was no longer between Arnoul and Gerbert (if it ever was), but a contest between king and pope, bishop and monk.

King Hugh Capet supported Gerbert.

Pope John XV supported Arnoul.

The French bishops supported Gerbert (and King Hugh) against the pope, whom they pointedly addressed as "the bishop of Rome." They considered him only "first among equals," not of superior rank to themselves.

The French monks, led by Abbot Abbo, supported Arnoul (and the pope) against the French bishops (and King Hugh). Their recalcitrance was so extreme—and even violent—that the bishops excommunicated the entire monastery of Fleury, and King Hugh called Abbo to task for fomenting French monks to riot against their lords.

The teenaged Emperor Otto III supported the pope. Gerbert had fallen from favor at the imperial court when Otto's mother Theophanu died in 991 and his grandmother, the aging Empress Adelaide, became regent. Adelaide had never forgiven Gerbert for his arrogant letter from Bobbio. He was also tainted by his close association with Theophanu, "that Greek woman" whom Adelaide had always despised. Abbot Odilo of Cluny, taking Adelaide's side, wrote, "While that Greek empress could be quite helpful and pleasant to herself and others, things were different where her august mother-in-law was concerned." He claimed that Theophanu once boasted, "gesturing with her hand as she did so, 'If I live another year, Adelaide's power in this world will be small enough to fit in the palm of my hand.'" "Divine judgment," Abbot Odilo continued, "guaranteed that her ill-considered words would come true: Before four weeks had passed, the Greek empress passed away from the light of this world, while the august Adelaide remained behind, safe and sound."

The pope, with the young emperor behind him, felt strong enough to challenge King Hugh and the French bishops directly. He used his direst weapon: He excommunicated Gerbert.

It was not as terrible as it sounds. As a young monk at Aurillac, remember, Gerbert had faced excommunication if he was proud, haughty, angry, jealous, or begrudging, or—and this was his current sin—he did not defer to his elders. The pope considered himself "elder" to any bishop or archbishop. The council of French bishops insisted he was not—he was just a bishop himself, no more senior than any other, and so could not excommunicate Gerbert unless they concurred—which they did not. Excommunication was, in any event, a temporary chastisement. If Gerbert showed the necessary contrition—at Aurillac, that meant prostrating himself before the chapel door, face to the ground—the pope would lift the sentence. Here, the pope demanded Gerbert give up the archbishopric of Reims; the king of France forbade it.

Yet, even with the king and the council of French bishops insisting Gerbert's excommunication was invalid, the pope's word carried weight. At Reims, Gerbert was shunned. As he wrote, "Not only my knights, but also the clerics conspired that no one would eat with me, no one assist in the sacraments. I keep silent about the vilification and contemptuousness." He appealed the pope's sentence. "The deepest grief overwhelms me as I learn that I have been removed from the fellowship of your very sacred apostolate," Gerbert wrote to the pope.

Yet he would not give up. In a letter to the French queen, he argued that the decision of a council of many bishops overruled the pronouncements of a single bishop—even if that one were the bishop of Rome. He wrote, "The church which I was charged with governing by the decision of the council of bishops I am unwilling to relinquish except by such judgment."

In February 996, he learned that his enemy, Abbot Abbo of Fleury, was on his way to Rome to see that Gerbert was finally, permanently, evicted from Reims. Gerbert, writing to his friend and solace, Constantine, said, "I am greatly amazed at the mission of venerable Abbo. . . . All of these things are not troubles, but the beginnings of troubles. Greater is his complaint and what he seeks than am I, who am humble and of little account." Abbo was seeking no less than to nullify the power of all bishops, to completely reorganize the hierarchy of the Church and make the pope's word supreme. "Even the kings themselves will appear as sinners," if Abbo wins, Gerbert said. "Let no one be pleased by the shattering of something while he himself remains unharmed."

Deciding his only hope was with the empire, Gerbert left Reims. He met Otto III's court at Ingelheim ten days later and traveled south with the young king of Germany to Rome in hopes of countering Abbo's accusations. Before they reached the city, Pope John XV died of fever. Things would turn out much differently—and much better—than Gerbert could have imagined.

CHAPTER XI

The Legend of the Last Emperor

King of Germany since age three, Otto III had gone to war against the Slavs at six, carried along as a sacred object to inspire his soldiers. On the battlefield, he received the homage of the Polish prince, who gave him a camel. He led armies at age twelve. Upon Empress Theophanu's death in 991, the hordes of Gog and Magog had descended on Germany. Otto marched three times against the Slavs in 993, then turned to face a Viking attack in 994. In August 995, his war-leader, Henry the Quarreler, died. Nonetheless, when Pope John XV called for his aid against the Crescentians, the fifteen-year-old emperor bravely marched on Rome. He left his grandmother, Adelaide—now no longer regent—in charge of Germany. When he had come of age, at fourteen, Otto had begun to question her policies. Meeting Gerbert at Ingelheim, he remembered him as his father's tutor and his mother's spy. He took no notice of Gerbert's excommunication. As they traveled south to Rome, the two became fast friends.

Otto III would be King Arthur to Gerbert's Merlin or, in the analogy of the day, Alexander the Great to Gerbert's Aristotle. Son of a Saxon king and a Byzantine princess, Otto was born to reestablish the Roman Empire, to rejoin East and West into one great Christian kingdom, from Constantinople to the islands of Britain and beyond. He was, at the very least, a new Charlemagne in the making.

Symbol of the Last Emperor, this grand processional cross of gold and gemstones on copper may have been a gift from Otto III to Charlemagne's cathedral at Aachen, where it remains. At its center is a cameo of Caesar Augustus, representing empire; on its back is an etching of Christ crucified—one of the earliest images of the suffering, human Jesus, instead of the all-powerful King of Heaven.

His father and grandfather had dreamed that dream, of a Christian empire spanning the cosmos. A legend popular in Byzantium during his mother's childhood, and which quickly spread through the West, foretold "a king of the Greeks and Romans" who would call all pagans to Christ. He would defeat the forces of Gog and Magog, loosed from the North. "Then the Earth will sit in peace and tranquility such as has never been seen nor ever will be any more, since it is the final peace at the end of time." The Last Emperor would travel to Jerusalem and hand his crown to Christ himself.

It was quite a burden for a sixteen-year-old. Otto's official documents betray a sense of sadness and urgency and of the crushing weight of duty. His strangely intense spirituality—when upset, he fled the palace, barefoot, in rags, and walked to a monastery—reveals a mind under heavy stress.

Otto was precocious, fanciful, impatient, rash, severe in judgment, and quick to repent. From his father's side, he inherited "fiery glowing eyes which sent forth a gleam like a flash of lightning" and the conviction that he, a king, could divine God's will. From his mother's side—from his Greek blood—came all his faults, it was said. He kept unkingly friendships. A boy named Tammo was "so intimate and dear to the king," says a medieval writer, "that they wore each other's clothing and often used a single spoon when eating." In an age when kings were walled off alone, austere, stern, and unapproachable, Otto opened himself to friendship: Like Gerbert, he wanted to be loved. Unlike his father, he had the mind of a scholar. He was delighted by Gerbert's learning, and immediately engaged him as his secretary and counselor. He agreed to back Gerbert's case before the pope and have his excommunication, at least, lifted.

His chance to help his new friend came sooner than expected. Celebrating Easter in Pavia, Otto learned of the pope's death. Asserting his right as emperor (though he had not yet been crowned) to choose the pope, Otto sent his cousin Bruno off to Rome to take John's place. Only twenty-five, Bruno was an ordained priest and well-enough educated to qualify as a bishop—even the bishop of Rome. A sizable army accompanied him, to make sure the fractious Romans accepted Otto's choice. On May 3, 996, Bruno was elected Pope Gregory V.

Eighteen days later, the new pope crowned his cousin Holy Roman Emperor in a lavish ceremony at Saint-Peter's. Otto wore a mantle embroidered with scenes of the Apocalypse in gold. Other details of his dress can be guessed from manuscript illuminations depicting him on his throne (see Plate 8). His mantle is clasped at the shoulder with a heavy round brooch bearing a blue gemstone surrounded by pearls. Under the mantle (green in one painting, red in another), he wore a long-sleeved tunic, stiff with gold bands of embroidery and glittering with gems at placket, hem, collar, elbow, and cuff. Beneath it, a voluminous robe of royal blue covers his high black boots. He perches on a magnificent throne—pink marble with arms carved into animal heads—and bears three symbols of rule: a crown, encrusted with gems and rising to three crosses; a scepter, topped by a golden eagle; and the orb of the world, a golden globe marked with a silver cross.

Acting as Otto's secretary, Gerbert sent notice of the coronation to the aging Empress Adelaide. It must have amused him to write, in the imperial "we": "Because the Divinity, in accordance with your wishes and desires, auspiciously conferred upon us the rights of empire, we do, indeed, adore Divine Providence and render true thanks to you."

It was Otto's first time in Rome. He would not have noticed, in May, the mosquito-infested swamps that flanked the Tiber from the city to the sea. He may not have known the medieval poem about the notorious air of Rome that brought so many deadly summer fevers: "The sluggish earth reeks, and fetid water lies in the swampy lakes; foul vapors slowly rise from the rotting marshes." He could not know that nearly every northern king or bishop who stayed in Rome over the summer inevitably died—or watched his partisans die—of fever, probably malaria: The connection between mosquitoes and malaria was not made until the late nineteenth century; that some genes conferred protection against the disease was not known until the twentieth century.

Instead, Otto would have marveled at the enormous ruins: the Colosseum, the triumphal arches, the aqueducts (some still in use), the baths and palaces, and the brick forums, with little lean-tos scabbed onto their walls. Smoke rose from furnaces that burned marble statues to make lime for mortar. The Pantheon, stripped of its marble cladding, was now a church, its famous oculus looking down on an altar to the Virgin Mary. Trajan's column, with its lively carvings of men and beasts spiraling up to a lookout point, was owned by monks who charged a fee to climb its internal stairway and gaze out at the view.

The greatest city in the world had long ago been reduced to squalor. Rome, *caput mundi*, head of the world, wrote Alcuin, Charlemagne's schoolmaster, in about 800: "Golden Rome, there remains to you now only a great mass of cruel ruins." And yet, its legend remained. "Nothing is equal to thee, O Rome," wrote the archbishop of Tours just after the year 1000, "even though your ruin is almost total; your ruins speak more eloquently than your former greatness."

A walled city built for a million, Rome now held a scant 50,000. Vast zones of the ancient city were overgrown with weeds. Vineyards and olive groves filled the valleys. Goats grazed the slopes. Monasteries crowned the seven hills. Most of the population clustered in the bend of the Tiber: On one side was the merchants' quarter; across the bridge was the Leonine City. Pope Leo IV had built the wall enclosing Saint-Peter's basilica after the center of Christendom was sacked in the 800s by 10,000 Sicilian Muslims. Having seen the Archangel Michael alight atop Hadrian's tomb, the pope reinforced the structure with battlements and called it the Castel Sant'Angelo.

Once inside its gates, a pilgrim could find anything he needed. Shops and foodstalls, stables and moneylenders; hostels organized by country and endowed by kings; peddlers selling pilgrim's badges, holy oil, candles, and religious icons; fountains and bathhouses; cells for hermits—they crowded up to the very doors of the huge basilica, its gigantic marble columns funneling the faithful to the tomb of the apostle, the light from oil lamps and candles glinting off the glass mosaics on the walls, off the paintings and frescoes covering every surface. From every side came the chanting of canons, the crying of hawkers, the pleas of beggars, and the ringing of bells.

A day's walk away was the second most holy spot in Rome: the basilica of the tomb of Saint Paul. Between Saint-Peter's and Saint-Paul's was the Lateran, the administrative seat of the pope, a palace rich in marble arches and glittering with mosaics illustrating the lives of the apostles. The famous statue of Lupa, the wolf-mother of Rome, sat in one courtyard, the equestrian bronze of Marcus Aurelius (then thought to be the Christian Emperor Constantine) in another. A white marble staircase joined two luxurious feast halls graced with columns of red porphyry, the royal stone. The larger of the two halls was 220 feet long and 50 feet wide; at its center was a marble fountain surrounded by divans on which the pope's guests could recline to eat like true Romans. Two fortified towers adjoining the palace provided the pope with a safe haven and a good place from which to study the stars, if he was so inclined. Across the cloister was the basilica of Saint John, housing the official throne of the pope. Palace and church were enveloped in a maze of houses and dormitories and offices and cells for the various monks, canons, deacons, subdeacons, cardinal-deacons, cardinal-priests, and cardinal-bishops, the noblemen, soldiers, artisans, bakers, butchers, brewers, and merchants who supported the work of the bishop of Rome, servant of the servants of God.

The "servant" now enjoying such luxury and power was Otto's twenty-five-year-old cousin, Bruno. As Pope Gregory V, Bruno immediately convened a church council, which Emperor Otto III attended. Gerbert was also invited, since the question of Reims was the chief problem to be discussed. Pope Gregory did not take Gerbert's part—to the astonishment of Gerbert's new friend. Otto had expected his cousin to put an end, once and for all, to the recurring power struggles between emperor and pope by kowtowing to his every wish. Gregory had other ideas. He gave Gerbert no clear answer.

He was slightly more disposed toward a second petitioner, another of Otto's new friends, Bishop Adalbert of Prague. Where Gerbert charmed Otto with his wide knowledge, Adalbert enthralled the young emperor by purity of spirit.

Adalbert was beautiful and noble—the picture of a saint. He had studied at Magdeburg under Otric, the schoolmaster Gerbert had so soundly defeated in debate at Ravenna. He became bishop of Prague, but had soon fled his post. He was being suffocated by sin, he had complained to Pope John XV: In Prague he had found men with two wives, priests living with women, Christian captives sold to Jews, while the duke had warred against the Poles, a Christian people, and made deals with the pagan Slavs. Pope John had sent Adalbert on pilgrimage to Jerusalem, but he made it no farther than the cell of Nilus the Hermit, who convinced a Roman monastery to admit him. The pope had ordered Adalbert out of the cloister and back to his post at Prague. Adalbert fled again to the monastery, where Otto III met him. They talked "day and night," a medieval source says, suggesting that something a little improper was going on between these two noble young men, both called to rule, to conquer and convert the heathens, though both preferred the peace of the cloister, where the only difficulties to overcome were bound between the covers of books.

At the council, Pope Gregory ruled that Adalbert must return to Prague and resume his duties as bishop, just as Pope John had decreed. In private, he agreed that Adalbert could perhaps risk a brief mission to the pagans in Prussia. But all Otto heard was that, once again, his cousin had gone against his wishes. Otto was annoyed. He determined to teach his cousin a lesson.

It must have given Gerbert great joy to write to Pope Gregory, in Otto's name, a few months later, when the bad air of Rome had finally come to the young emperor's attention: "I am overcome by vehement grief because the unseasonable weather has prevented me from satisfying your desires. For I am urged on by an affectionate piety towards you, but the necessity of nature, which restricts everything by its own laws, puts into opposition the quality of the Italian climate and the frailty of my body." Otto was returning to Germany for the sake of his health, of course—not to punish Gregory for being too independent. Yet a German pope would not last long without a German army in Rome, and both pope and emperor knew it. Gregory was expected to beg Otto to stay. He did not.

Otto returned to Germany, Gerbert to Reims, Adalbert to Prague. A month after they left, Pope Gregory was chased out of Rome by the Crescentians. In early 997, he was replaced by an antipope. He would not regain his seat until Otto and his army returned in 998.

Gerbert, still excommunicated, reached Reims in late summer 996; by the fall, he knew he could not stay. After reigning less than ten years, Hugh Capet, age fifty-five, died of smallpox. His son, Robert the Pious, succeeded him. Twenty-four years old, Robert was a great warrior, tall, fair-haired, with a long nose and a "gentle" mouth, "affable and pleasant," "prudent and erudite," said his contemporaries, a "man always ready for peace and respectful to all." Gerbert had taught him well at Reims: He composed hymns, read holy works, and carried his bookchests wherever he went. He was generous to churches and monasteries, fed and clothed the poor, and even—like Count Gerald the Good of Aurillac—healed the sick with the water that had washed his hands. He would be king of France for the next thirty-five years.

He and Gerbert immediately had a falling out over Bertha, the beautiful countess of Blois. The countess had just been widowed, and King Robert, who had long admired her, proposed marriage. Unfortunately, he was already married. Having failed to procure a Byzantine princess, Hugh Capet had wed his son at sixteen to Rozala, daughter of a king of Italy, widow (and mother) of a count of Flanders, and somewhat over thirty. Robert called her his "old Italian" and set her aside. Rozala's kinsmen did not recognize the divorce, since Robert refused to return her dowry, so the king was still, technically, married. There was another problem: Robert and Bertha were second cousins. The Church at the time considered such a marriage incestuous. And a third problem: Robert was already the godfather of one of Bertha's five children. This spiritual kinship, by itself, was enough to rule out marriage. Gerbert—whom Robert still considered the archbishop of Reims, in spite of the pope's sentence of excommunication—refused to perform the wedding. His chances of attaining Robert's friendship—or favor—vanished.

The king found a more pliant archbishop at Tours and married Bertha there. Pope Gregory promptly excommunicated both the king and the archbishop of Tours.

Then Abbo of Fleury, Gerbert's eternal enemy, made a suggestion to the king: Give the pope his choice for archbishop of Reims, and perhaps he will withdraw his objections to your queen. Robert liked the idea. Abbo set off again for Rome. Finding, not Pope Gregory, but an antipope in Saint Peter's chair, Abbo wandered "through deep valleys and across precipitous mountains," as he later wrote. He finally located Gregory in Spoleto, where he had hidden himself for fear of the Roman nobles. What deal Abbo made we don't exactly know. Back in France, he wrote to Pope Gregory, "Venerable Father, I conveyed your views to

King Robert faithfully and simply, as you bade me, fearing no enmity from the king through loyalty to you; I added nothing, cut out nothing, changed nothing. Sure evidence of this lies in the fact that Arnoul, freed from captivity, is now archbishop of Reims." The pope had gotten his choice for archbishop, but he still did not bless Robert's marriage to Bertha. The king remained excommunicated as long as Gregory V remained pope.

If anyone benefited from this deal, it was Abbo: He had triumphed over Gerbert at last. Better yet, he had triumphed over the bishop of Orleans. According to a new charter that the pope conferred on Fleury, no bishop could say Mass at the monastery—or even enter its grounds—without Abbo's invitation. If accused of wrongdoing, the abbot could be judged only by the pope himself. The charter's flowery phrases matched a bull attributed to Pope Gregory IV confirming similar privileges on Fleury in the early 800s—except that this "papal bull" was a forgery, written by Abbo himself. The abbot was not a little proud of himself. "I am more powerful than our lord the king of the Franks in these lands where no one fears his rule," he bragged.

Gerbert did not wait for Arnoul to show up at Reims to take his place. He copied some of his letters and *The Acts of Saint-Basle* and sent them off to his friend, Constantine, for safekeeping. He politely rejected the latest student sent for training, for fear of "evil befalling the boy because of the difficult times." To Bishop Arnulf of Orleans, still a true friend, he entrusted the property he owned, both houses (with their furnishings) and churches. Many of his books he boxed up to take with him, particularly the lavish copy of Boethius's *On Arithmetic*, written in gold and silver inks on purple parchment, which he would give Otto III as a gift. No one knows what happened to his abacus board, his celestial spheres (including the one he was making for Remi), or his monochord; his organs were lost in Italy.

He penned a few last letters before leaving France for good. To the dowager-queen, King Robert's mother, who had invited him to join the

French court, he wrote of the pope's decree: "The too wicked and al-most incredible report has affected me with such grief that I have almost lost the light of my eyes in weeping, but when you order me to come to you to offer consolation—an excellent idea, indeed—you command the impossible. For my days have passed away, O sweet and glorious lady. Old age threatens me with my last day." He was fifty years old and, he said, in poor health. "Pleurisy fills my sides; my ears ring; my eyes fill with water; continual pains jab my whole body. This whole year has seen me lying ill in bed, and now, though scarce out of bed, I have suffered a relapse, and am seized by chills and fevers on alternate days." This is a rather exact description of tertian fever—a kind of malaria, which he could have caught in Rome. Yet here it is merely an excuse. Gerbert knew he had too many enemies in the French court, and the king's mother was too weak to protect him. Poor health or no, he fled instead to Germany.

Reaching Emperor Otto's side at Magdeburg in June 997, he wrote the queen again, refusing rather more strongly to return to the French court. "It is inconceivable that my return would be without danger to my head. Even if you will not take notice of it, I ought not to doubt that this is so." He would no longer contest the pope's decision, nor "wrack your church by schism. Indeed, I am acutely aware of the cun-ning of these wicked men" (he doesn't need to name Abbo and Arnoul), "but if so decreed I will defend the unity of the church against all schisms by my death." He would never return to Reims.

In Magdeburg, at Otto's court, he worked his usual magic. He organ-ized a grand debate on rhetoric, which he wrote up later as a treatise, *On Reason and the Uses of Reason*. He built Otto the astronomical in-strument Thietmar of Merseburg called an *horologium*. "Astrolabe" is one possible translation, but whatever Gerbert made, it was a new and delightful toy for the court. And Otto was in dire need of diversion. There was trouble to the south in Rome. The Slavs were attacking from the east. And his beloved Adalbert of Prague had just been martyred by the Prussians in the north. According to the account by Bruno of Querfurt (who himself would be martyred by Prussians in 1009), the beautiful Adalbert stayed silent while his captors dragged him one Friday, chained, to the top of a hill and pierced him with seven lances. He fell with his arms outstretched in the shape of a cross. As he died and his soul fled, his fetters burst open.

Otto built a church in Adalbert's honor, commissioned jeweled crosses, petitioned to have his friend sainted, and planned a pilgrimage to his tomb. Rewarding Gerbert for his friendship, he gave him a magnificent German estate called Sasbach. Then Otto returned to the war against the Slavs.

Writing from the estate, Gerbert guided and supported Otto with wit, fondness, and fatherly good sense. "I am especially concerned to inform you of the wails and groans of brother W," he wrote. "That noble man bewails the fact that his brother is being destroyed by hunger near Gorze, regardless of his rank and that of his family, and this leads to everlasting disgrace." Captors must be particularly mindful of the humane treatment of their captives, he reminded the young emperor, "lest, after regaining liberty, the captives . . . injure the captors or their friends."

Upon hearing of Otto's Slavic victories, he said, "You could make known to us, solicitous in your behalf, nothing sweeter than the great renown of your empire.' . . . May the last number of the abacus be the length of your life."

When the previous owner of Sasbach forcibly reclaimed the estate, Gerbert (master of rhetoric that he was) let loose a stream of petulance at his young patron:

> I know that I have and do offend the Divinity in much. But I do not know by what contradiction I am said to have offended you. . . . Would that either I might never have been permitted to receive such gifts, offered with glory by your generosity, or, once received, I had not lost them in such confusion. What should I think of this?
>
> What you most certainly gave, you were either able to give, or you were not. If you were not able to, why did you pretend to be able to?

If you were able to, what unknown, what emperor without name, commands our emperor, so noted and famous throughout the world? In what shadows is that scoundrel lurking? Let him come forth into the light and be crucified so that our Caesar may freely rule.

It was not Sasbach, however, that Gerbert wanted. He reminded Otto that he had served his father and grandfather and had offered the young emperor himself "the most incorruptible fidelity" since his birth: "For your safety have I exposed my person, however small, to raging kings and frenzied people. Through wilderness and solitude, beaten by the assaults and attacks of thieves, tortured by hunger and thirst, by cold and heat, in all disturbances I stood firm so that I chose death rather than not to see the son of Caesar, then a captive, on the throne." So to the clinch: "I have seen him rule, and I have rejoiced. Would that I might rejoice to the end and finish my days with you in peace." He enclosed the purple and gold book on arithmetic.

At the same time as he was covering Otto with guilt, Gerbert wrote to a fellow bishop that he was doing quite well, thanks. "I am confident that I am being freed from Ur of the Chaldees" (Gerbert's name for Hell—and Reims). He was not really upset over losing Sasbach. "This event will accomplish what I have always wished, have always hoped for, by making me his"—Otto's—"inseparable companion so that we can guide the exalted empire for him. What, therefore, could be sweeter? What more outstanding?"

Gerbert got his wish. The letter came in October 997 from Aachen— intimate, playful, mocking, and imperious, written by the hand of Otto himself: "We wish to attach to our person the excellence of your very loving self," Otto wrote. Why? Because "your philosophical knowledge has always been for Our Simplicity an authority not to be scorned." He ordered Gerbert immediately to his side so that "your expert knowledge may be zealous in correcting us, though not more than usual, unlearned and badly educated as we are, both in writing and speaking, and that with respect to the commonwealth you may offer advice of the highest trustworthiness." Furthermore, he added, "We desire you to show your aversion to Saxon ignorance by not refusing this suggestion of our wishes, but even more we desire you to stimulate Our Greek Subtlety to zeal for study, because if there is anyone who will arouse it, he will find some shred of the diligence of the Greeks in it." As a postscript, he added, "Pray explain to us the book on arithmetic."

That peculiar reference to "Saxon ignorance"—other translations are "Saxon boorishness" and "Saxon rusticity"—has opened Otto to accusations that he was un-German, even anti-German. According to some (mostly German) historians, he was a failure who squandered his potential and died young *because he turned his back on Saxony* to chase imperial butterflies. And who put the fantasies of empire in the silly youth's mind? Gerbert.

It is true that Gerbert dreamed of restoring the empire. As a young man in Catalonia, passing through the church of Elne with Miro Bonfill, Gerbert had carved his name in the shape of two symbols: cross and chrismon, church and empire. The elaborate acrostics in his *Carmen Figuratum* reduce to the same idea: Otto II as the new Constantine or Charlemagne. The treason of Archbishop Adalbero and the crowning of Hugh Capet served the goal of empire.

Gerbert didn't hide his aspirations. "Ours, ours is the Roman Empire," he gushed to Otto III, after agreeing (of course!) to be his teacher and counselor. "Italy, fertile in fruits, Lorraine and Germany, fertile in men, offer their resources, and even the strong kingdoms of the Slavs are not lacking to us" (Otto had won the latest battle). "Our august emperor of the Romans art thou, Caesar, who, sprung from the noblest blood of the Greeks, surpass the Greeks in empire and govern the Romans by hereditary right, but both you surpass in genius and eloquence."

But imperial dreams were not the snares Gerbert used to catch young Otto. Those dreams the boy had inherited from Theophanu and Otto II, Adelaide and Otto I: Gerbert merely shared them. Gerbert's snares were "the treasures of Greek and Roman wisdom" that he offered to impart to the quick-thinking young emperor. Otto wanted more astrolabes (or other instruments) to study the stars, "zealous" correction in writing and speaking, an organ like his father's (Gerbert's official title at court would be "musician of the chapel"), and help fathoming Boethius's *On Arithmetic*.

Accepting Otto's invitation, Gerbert complimented him: "For, unless you were not firmly convinced that the power of numbers contained both the origins of all things in itself and explained all from itself, you would not be hastening to a full and perfect knowledge of them with such zeal. Furthermore, unless you were embracing the seriousness of moral philosophy, humility, the guardian of all virtues, would not thus be impressed upon your words. Not silent, moreover, is the subtlety of a mind conscious of itself."

Or, as he phrased it in his treatise on reason, written soon after at the emperor's request, "In the same way that reason separates us from all the other animals who are incapable of reasoning, it is *the use of reason* that makes us different from those animals" (and he doubtless had certain men in mind here) "who do not use reason."

It was the curriculum of the cathedral school of Reims that ensnared Otto. Only by obtaining wisdom, by endeavoring to understand "all things human and divine," could Otto live up to his duties as the Last Emperor.

CHAPTER XII

The Pope of the Year 1000

Otto, an emperor at sixteen, did not have much time to learn mathematics and music or study the stars. But from 997 on, he kept Gerbert close. He trusted his advice. He leaned on Gerbert's eloquence to restrain "the minds of angry persons from violence," though it didn't always work. He had him write letters and erect siege engines and direct learned debates at court. And when Saint Peter's chair again came empty, in 999, the wise Gerbert was Otto's obvious pick.

For Gerbert, it was a mixed blessing. He had seen the papacy first-hand since following Otto to Rome, and found little that he desired. It was Bobbio squared, Reims to the *n*th: petty politics and feuds unending, with no time to contemplate the timeless, to explore God's creation through number, measure, and weight. "What, therefore, am I, a sinner, doing here?" he might have thought.

It was dangerous to be pope in the tenth century. Gerbert's predecessor, Pope Gregory V—imperial cousin, son of a duke, great-grandson of Otto the Great—died of fever, but his life had been under threat for the past three years. After Otto III took his army back to Germany in the summer of 996, Gregory lasted only four months before he was chased out of Rome by a mob. The antipope who replaced him was John Philagathos, abbot of Nonantola, archbishop of Piacenza, chancellor of

Italy, and "dear companion" of Empress Theophanu (some said even her lover). Philagathos's fate, which Gerbert also witnessed, was worse, much worse, than Gregory's.

Born in Calabria, the Greek-speaking part of Italy, Philagathos had joined Theophanu's court before Otto's birth. Some sources say he was Otto's godfather, others that he tutored the boy in Greek. In 994, Otto sent him to Constantinople to find him a royal Byzantine bride, and so he was not at hand in 996 when John XV died and Otto appointed his cousin as pope. Returning less than a year later, Philagathos felt unjustly overlooked. His traveling companion, Leo of Synada, whom the Byzantine emperor had sent to continue the marriage negotiations, agreed. Meeting with the Prince of Rome, Crescentius of the Marble Horse, the two ambassadors urged him to appoint a new pope. So he did. Gregory was chased out of town in September 996. Philagathos was acclaimed John XVI by the citizens and senate of Rome (a detail missing in Gregory's election) and anointed in February 997. He would last until Otto arrived with his army.

Leo, the archbishop of Synada, was a wily trickster, as his letters show, quite willing to play on Philagathos's hurt pride. "You will laugh out loud to hear that I have elected as pope that Philagathos whom I should have smothered," he wrote to one correspondent in Constantinople. To others: "I announce to you the news that Philagathos is pope. To whom should we attribute this development? . . . Rome is at the feet of our emperor, thanks be to God. And while I contributed, God directed the heart of Crescentius. . . . I saw Rome, and I took bold action. If the emperor is satisfied, so much the better. If he is not, you'll take care of it. . . . But of this bastard Philagathos, may I see him brought low just as I saw him raised up."

By letter, Otto demanded that Philagathos explain himself. The royal messengers were intercepted by Crescentius, however, and Otto received no reply.

Nilus the Hermit warned Philagathos to submit to the emperor and beg his forgiveness; Philagathos agreed to do so, but his humble letter to Otto also went astray.

Furious, Otto marched on Rome. His army of Germans, led by Gregory V's father, cowed the city into surrender after one skirmish. Philagathos fled. Crescentius walled himself up in the Castel Sant'Angelo and held out for two months, until Otto's siege engines (possibly designed by Gerbert) broke through. Crescentius of the Marble Horse was beheaded and hanged by the feet from the castle walls alongside twelve of his companions.

Philagathos was captured by Berthold, count of Breisgau. "Fearing that if they sent him to the emperor, he might depart unpunished," say the *Annals of Quedlinburg*, Gregory V's German partisans took matters into their own hands. Leo of Synada gleefully tells the story: "Now you are going to laugh, a big, broad laugh, my dear heart and soul," he begins. Philagathos, whom Leo clearly never liked, has fallen:

And why shouldn't I tell you, brother, openly how he fell? Well, first, the Church of the West dealt him anathema; then his eyes were gouged out; third, his nose, and fourth, his lip, and fifth, that tongue of his which prattled so many and such unspeakable words, one by one, were all cut from his face. Item six: He rode like a conqueror in procession, grave and solemn on a miserable little donkey, hanging on to its tail. . . . Then they put on him the priestly vestments, back to front, the wrong way round; and then they stripped them off again. After this he was pulled along the church, right through it, and out by the front portico to the court of the fountain. Finally, for his refreshment, they threw him into prison.

Nilus the Hermit was outraged. He marched on Rome protesting Philagathos's treatment. Ninety years old and skeleton-thin, this ascetic was revered as a living saint. Otto and Gregory begged to kiss his hands. He refused. He called down the wrath of God on them. No one knows who, if anyone, had approved the antipope's torture and mutilation, but the eighteen-year-old Otto was distraught. He had known Philagathos and loved him. He was ultimately responsible for the horror—for not upholding the kingly, and Christian, principle of mercy.

Through an intermediary—probably Gerbert—Otto begged Nilus to assign him penance, but the hermit would not hear him. "If you do not forgive him whom God has delivered up into your hands, neither will the heavenly father forgive you your sins," Nilus replied, and promptly left Rome.

When Pope Gregory died of malaria a year later, at the age of twenty-eight, Nilus's biographer claimed he had been poisoned—and that his eyes were ripped out beforehand in revenge for Philagathos.

Otto, having decided upon his own penance, was making a barefoot pilgrimage to Nilus's hermitage south of Rome when news came of his cousin's sudden death. Strangely, one of the first things Otto did was to heap honors on Count Berthold of Breisgau, who had captured and mutilated Philagathos. Berthold was the first German layman allowed to hold a market on his own estate, to mint coins, and to collect tolls. He was the emperor's stand-in, entrusted with the golden crozier, when Otto's sister Adelaide was installed as abbess of Quedlinburg following the death of their Aunt Matilda. These rewards show that Otto, though he regretted Berthold's means, understood his ends. The emperor's power in Rome—and his right to choose the pope—had to be upheld at all costs.

Then Otto appointed Gerbert pope.

Pope Gregory had made Gerbert the archbishop of Ravenna in April 998, in that harried time between Crescentius's beheading and Philagathos's mutilation. It was the only way to appease his cousin the emperor, since Gregory no longer had the power to restore Gerbert to Reims. Giving Gerbert the archbishopric of Ravenna—second only to Rome among the churches of Italy—allowed pope and emperor to call a truce; it also meant Gerbert's sentence of excommunication was lifted.

It came just in time. It might have been awkward if the new pope could not say Mass. Yet doubtless Otto would have found a way around it.

Many of Gerbert's successors were even less likely popes: John XVII was married with three children (he lasted less than six months). John XVIII (1003–1009) was the bastard son of a priest. Sergius IV (1009–1012), known as Peter Pig's Snout, the son of a shoemaker, was indeed a bishop; he had been Gerbert's papal librarian. But Benedict VIII (1012–1024) was a layman, son of the count of Tusculum. Benedict ruled Rome alongside his brother Romanus, who was consul and senator and then pope, in his turn, as John XIX (1024–1032). Neither Benedict nor John were churchmen. Both were elevated to the priesthood *after* becoming pope, so that they could perform their duties on the eighty-five days a year when the pope made his grand procession through the streets of Rome, trailed by clergy, commoners, and pilgrims from every Christian land, to preside at one or another of the many churches in the Holy City. Eighteen times a year, the pope was solemnly crowned during Mass.

Such rituals were meant to reinforce his rank as spiritual lord of the city—and his superiority over other bishops, who weren't nearly so splendid. Yet the papacy in Gerbert's time was just one more center of power. The bishop of Rome's authority over other bishops and archbishops—as the long fight over Reims illustrates—depended on the situation, and especially on the backing of kings, emperors, and Roman nobles. Over the next century, thanks to reforms begun by Abbot Abbo of Fleury and (oddly) Gerbert himself, among others, the power structure of the Church would be completely transformed. From 1099 through the end of the Middle Ages, the pope would represent a spiritual and political force that kings and bishops could ignore only at great peril.

The peril in Gerbert's day (as he knew well) was not so very great. He had been excommunicated by the pope himself and was still welcomed at both the French and German courts. King Robert the Pious, himself excommunicated, shrugged it off for five years. (When it became clear that his beloved Bertha was not going to produce an heir, he demoted her from queen to mistress, took a new, proper bride, and was admitted once again into Christian fellowship by Gerbert, his former schoolmaster.)

In the balance of power between emperor and pope, the pope was decidedly weaker at that time. Faced with an offended emperor, as Philagathos and others before him had learned, the pope was powerless.

Gerbert had no intention of offending Otto, or even of thwarting him, as Otto's cousin Gregory had. Gerbert saw no point to a pope and an emperor working at cross purposes. He had seen church and state cooperate in Catalonia, so many years ago. He believed—and had taught Otto—that kings should be philosophers, ruling through reason and law. Both he and Otto agreed on the necessary structure and hierarchy of empire, led by an emperor who worked God's will on earth. Both believed in learning and logic as the antidotes to force and passion. Only when the kingdoms were at peace, the empire secure, they knew, could the Church prosper and fulfill its primary mission: worshipping God and his creation through the study of number, measure, and weight.

To signify his willing partnership with the emperor, Gerbert took the name Pope Sylvester II. It may have been Otto's choice, for the young emperor prized his gold-initialed copy of the life of Saint Sylvester. This Sylvester and Emperor Constantine had converted the Roman Empire to Christianity; the second Sylvester and Emperor Otto would renew that empire and extend its jurisdiction to the ends of the earth. *Renovatio Romani imperii*, read the new insignia with which Otto sealed his letters; it had been Charlemagne's seal as well: *To Renew the Roman Empire*.

"Ours, ours is the Roman Empire!" Gerbert had enthused, and on the eve of the year 1000 his wish seemed to be coming true. The brilliant young half-Saxon, half-Greek, known as the *mirabilia mundi*, the Wonder of the World, was matched with the Scientist Pope, the foremost intellectual of his day.

It was not to be. In the parlance of the time, it was not God's will. The empire of reason would last only three years. A fever claimed Otto's life in 1002. Gerbert died shortly afterward, his hopes shattered. For Europe, a great opportunity was missed.

The Scientist Pope accomplished nothing specifically scientific while in office. He did not install an astronomical observatory in one of the churches of Rome, for example, as Pope Clement XI did, seven centuries later. Thirty of Gerbert's official papal documents remain in full; another seventy to eighty existed as recently as the eighteenth century or are alluded to in other writings. They say next to nothing about science or mathematics. Adalbold of Liege, whom Gerbert, as archbishop of Ravenna, had taught to find the area of a triangle, asked him a question about the volume of a sphere, but we do not have his reply. Gerbert's epitaph for Pope Gregory is a chronogram, a puzzle concealing the number 999; to solve it, you replace each letter of the alphabet with a number, add them, then divide by the sum of the number of letters in each line. Even his favorite subject, books, Gerbert mentions only once, revealing the poor state of the papal library: "Concerning the matter on which you sought our advice, we have postponed answering you for the reason that we do not have the authority in books here in Rome. We remember having left in France those very books in which we read the particular opinion."

His most famous papal acts exist in legend—the documents are missing, but it was during his papacy that Christianity triumphed over paganism in Europe. Gog and Magog were defeated, not through battle, but by baptism. Gerbert wrote letters to Vladimir, prince of Kiev, and to Olaf Tryggvason, king of Norway, supporting their efforts to convert their countrymen (and commanding the Norsemen to cease using Viking runes and to write in Latin like civilized folk). He confirmed the ecclesiastical arrangements Pope Gregory had made with Boleslav Chobry, duke of Poland. To King Vajk of Hungary, baptized as Stephen, he sent a papal blessing and a royal crown (the very one, so it is said, now in the National Museum in Budapest), and he established the first Hungarian archbishopric at Gran. He sent another bishop (soon to be a martyr) to Prussia, and encouraged missions to the pagan Magyars, Liutizi, Pechenegs, and Swedes. By the time of Gerbert's death, the Roman Church stretched from Greenland east to the Black Sea, where it met its Byzantine counterpart, and talks were underway to reunite the Eastern and Western churches.

Most of Gerbert's acts as pope were more routine. He tidied up some practical matters that had bothered him throughout his career: Who should receive taxes, abbot or bishop? He substituted the French system of fiefdoms for the Italians' "little books" where he could, helping to codify the feudal system. He settled petty disagreements among the higher clergy—or tried to.

One feud that took up an inordinate amount of his time was instigated by Otto's older sister Sophie. Partly, it was a question of church structure. Who owned the convent of Gandersheim? Who made the decisions—and collected the income? Who judged its disputes, the pope or the German bishops? At bottom, though, the problem was personal: Who had the right to tell Princess Sophie what she could and could not do?

Sophie was Theophanu's second child. She had been made a nun at age ten or twelve, but had rebelled. She frequently left the convent of Gandersheim to rejoin the court. Her abbess's complaints were silenced by Willigis, archbishop of Mainz, who felt a princess should be given some latitude. In 996, age eighteen, Sophie traveled with Otto III and Gerbert to Rome and saw her brother crowned emperor by her cousin Pope Gregory. There her un-nunlike behavior was remarked on by Bernward, who as bishop of Hildesheim oversaw the convent of Gandersheim. Sophie took criticism badly—and she held a grudge. When it came time for a new church at Gandersheim to be consecrated, Sophie arranged for the more tolerant Archbishop Willigis to do the honors. Bishop Bernward, snubbed, objected. He arrived on the appointed day, but was met by armed guards. Having brought knights of his own, he persevered, and in spite of the nuns' attempts to stop him, held the ceremony. When he reached the offertory part of the Mass, the nuns threw their offerings at him and cursed him. Then they had Willigis reconsecrate the church the next Sunday. Bernward set off for Rome to complain.

Bernward is known to posterity as an extraordinary artist. He designed and cast the magnificent bronze pictorial doors, with their aggressive lion-head knockers and poignant scenes from Christ's life, that still adorn the church of Hildesheim. Inside the church is more of his work, including a seven-ton bronze pillar patterned on Trajan's column, with the Savior's life story spiraling in bas-relief up its curves. Two of his bronze candlesticks, each fourteen inches high, grace the altar: From lion-claw feet rise winged dragons bearing nude men; they fly up toward masses of foliage and flowers, from which peer devils, a snake, a lion, eagles, and angels, along with more nudes, eating grapes. Three salamanders hold the grease cup. Supposed to represent Nature striving to escape the darkness of evil and ascend toward the light of God, these candlesticks are a riot of imagination.

It was Bernward, as a boy, who tagged along with Thangmar, the schoolmaster of Hildesheim, as he ran errands, making up verses, arguing points of logic, and being read to as they rode along. Bernward had later been Willigis's student, but their friendship had dissolved over a quarrel. The emperor himself had been Bernward's student from age seven to thirteen. Otto felt such tenderness toward his old teacher that, when he heard of his coming to Rome, he rushed out of the palace to greet him. He pressed presents on the bishop: a splinter of the True Cross, a handsome onyx vase, bundles of spices and medicinal herbs, and the arm of Saint Timothy. He kept the bishop by his side as long as possible, and convinced Gerbert to confirm in writing Bernward's rights to Gandersheim (including the convent's taxes and tithes). When it came to a choice between his sister Sophie and Bernward, Otto chose his beloved teacher.

Gerbert sent a papal legate north. The pope's message gave total control of the rich convent to Bernward. Sophie's champion, Willigis of Mainz, was not impressed. He belittled the legate's fancy Roman saddle and ornate purple saddlecloth. He refused to read the pope's letter and, when it was read aloud to him in a gathering of churchmen, dismissed it as nonsense. Willigis had been an archbishop when Gerbert was a mere monk. He had been Otto III's guardian when the emperor was a toddler of three. For thirteen years, he had been the most influential churchman in Germany, adviser both to Empress Theophanu and to her mother-in-law, Empress Adelaide. He knew the German nobles were losing patience with Otto, who seemed to spend all his time nowadays in Italy, propping up the pope. When Bernward and the papal legate tried to enter the convent of Gandersheim, they were barred by a small army; they returned, defeated, to Rome.

The fight over Gandersheim dragged on. The German bishops met in a synod. Like the French bishops in the debate over Reims, they did not recognize the pope's authority. Nor were they impressed that Bernward had the emperor on his side. They ruled in favor of Willigis—and the princess Sophie. Again and again, Gerbert insisted that Gandersheim belonged to Bernward. Again and again the pope's pronouncements were ignored. Not until 1007, four years after Gerbert's death, was the ownership of Gandersheim decided—in favor of Bernward—by the forceful intervention of Otto's successor, King Henry II.

Gerbert was, however, able to settle two church disputes that had plagued him personally: those over Bobbio and Reims. He confirmed Petroald as abbot of Bobbio and—one suspects with great sarcastic glee—Arnoul as archbishop of Reims. He wrote to Arnoul, "To the apostolic pinnacle belongs the duty not only of advising sinners, but also of actually lifting up those who have lapsed." How good it must have felt to lord it over his rival. "Therefore, we have thought it fitting to assist you." Arnoul was justly deprived of his post, the pope explained, "because of certain excesses." But "because your abdication lacked the assent of Rome," Gerbert, "through the gift of Roman compassion," would reinstate him. In addition, no one (excepting the pope, who was doing exactly thus in his letter) "shall in any way presume to allege the crime of your abdication against you or to break out into words of reproach against you." Instead—here's the final twist of the knife—"everywhere our authority shall protect you, *even when a feeling of guilt shall seize your own conscience.*"

Only occasionally (as here) does Gerbert's voice shine through the papal mask. Another fine example is his rebuke of the bishop of Asti, who backed the rebellion of Margrave Arduin of Ivrea against Emperor Otto: "The whole world cannot endure the stench of your obscene infamy. . . . You prefer to putrefy midst your dung with the beasts of burden rather than to shine among the pillars of the Church." (In this case,

Gerbert might have been better off remembering his missteps at Bobbio and tempering his language; upon Otto's death, the margrave would become King of Italy and so Gerbert's overlord.)

Several of Gerbert's papal acts concerned Spain. Replacing charters that burned when al-Mansur the Victorious pillaged Barcelona and Girona in 985, Gerbert confirmed the rights that bishops and monks claimed to "manors, manses, manor houses, castles, cottages, vineyards, lands and various estates, cultivated and uncultivated, with their tenths and first fruits, male and female serfs and slaves and free farmers," churches with their tithes and offerings, hermits' cells, fortifications, walls, water rights, market tolls, court fees, pasturing fees, port dues and anchorage fees, and "the tax paid by the Jews."

And while Gerbert had, at Reims, championed a bishop's right to control the nearby monasteries and convents, as pope he saw things a little differently. He agreed with his enemy Abbo of Fleury that some monasteries, at least, should obey the pope alone. The issue to him was not monks' rights but papal power.

To Adelaide, new abbess of Quedlinburg and eldest sister of Otto III, for example, he declared that "this same place of Quedlinburg (located, it appears, on a mountain) may manage its own affairs" and be "immune from the yoke of obedience" to anyone but the pope. (Perhaps he should have tried this tack with Sophie at Gandersheim.)

The abbot of Fulda could be judged only by the pope—the same privilege Abbo of Fleury had acquired from Gerbert's predecessor. "No bishops, archbishops, nor, by chance, patriarchs, may celebrate the solemnities of the mass on an altar under your protection . . . but let the church of Fulda, always free and secure, zealously serve the Roman see alone." Above all, he warned, "No one shall carry off or give to anyone any of the revenues and incomes or tithes and other donations" to the church, as had happened to Gerbert at Bobbio.

The odd thing about the Fulda privilege, to modern eyes, is the date: "Given on the 31st of December . . . in the first year, with God's favor, of the pontificate of Lord Sylvester the Second, pope, with the pacific Otto III ruling, in the fourth year of his reign, the thirteenth year of the indiction." Indiction years were periods of fifteen years used in Rome for tax purposes. Christians counted indictions from September of the year the Emperor Constantine was converted, or A.D. 313, which makes this date without question December 31, 999.

On the very day the world was destined to end, Gerbert was bestowing a privilege upon the abbot of Fulda *and his successors, in perpetuity.*

Did he not fear the End of the World? No. The Terrors of the Year 1000 is a myth. The idea that all of Christendom quaked and cried in fear that the world would come to a crashing end precisely at midnight on December 31, 999, was made up by later historians hoping to brighten their own age by darkening the ones that came before. In no medieval text was the Apocalypse exclusively connected with the year 1000. The phrase "Clear signs announce the end of the world; the ruins multiply" had begun appearing on church charters in the seventh century and was still popular in 1035. Ralph the Bald recorded signs and wonders prophesying the End of the World from 956 until his death in 1046. Most medieval writers did not even use the Anno Domini system—and so did not know when the thousand years were ended and the devil would be loosed upon the world. Like Gerbert, in this example, they referred to the pontificate, the emperor's reign, or the indiction years. When they did use A.D., they disagreed on which dates were correct. Abbo of Fleury thought the real "Year 1000" since Christ's birth was our 979.

That said, the End of the World was never far from Gerbert's mind.

Otto III was crowned emperor wearing a robe embroidered with scenes of the Apocalypse. Elaborately illustrated copies of a commentary on Revelation by the eighth-century Spanish monk Beatus of Liebana were crafted in Catalonia while Gerbert was there. Otto himself commissioned a majesterial volume now known as the Bamberg Apocalypse; made in the monastery of Reichenau, it was not finished until after his death.

The Apocalypse, the end of the world we know and its transformation into the kingdom of Christ, was the ultimate goal of Christianity,

The Last Judgment from the Bamberg Apocalypse, a manuscript commissioned by Emperor Otto III and created at the monastery of Reichenau. The book was not quite finished when Otto died in 1002.

the concluding vision of the Bible. Without it the Church lost much of its clout. Who would abjure sin if there were no Judgment Day? At the same time, Christians like Gerbert did not only fear the Apocalypse, they desired it as an end to fear and suffering, the longed-for arrival of lasting peace. They hoped to be among the Chosen lifted by the hand of Christ straight into Heaven.

Otto III knew exactly the part he was to play in that scenario. He had read Adso of Montier-en-der's *Book of the Antichrist*. It was Adso who, in 954, linked the End of the World to the Legend of the Last

Emperor. Apocalypse could not come—never mind the year 1000—
"as long as the kings of the Franks who hold the empire by right shall
last." Only when the Last Emperor handed his crown to Christ upon
the Mount of Olives would the End be achieved. That emperor might
be Otto, as he hoped, or Otto's son—negotiations for a Byzantine bride
for the twenty-year-old emperor were continuing.

So while Gerbert plunged into the administrative morass of the
papacy, Otto made himself ready for his impending interview with
Christ. Hearing, as Thietmar of Merseburg records, "of the miracles
which God was performing through his beloved martyr Adalbert,"
Otto made a pilgrimage to Gniezno, Poland. He entered the city barefoot, wearing sackcloth, and "weeping profusely," prayed before Adalbert's tomb.

The Polish prince, Boleslav Chobry, arranged a great pageant of warriors and nobles wearing furs and fine cloth ("not linen or woolen," according to the twelfth-century chronicler known as the Anonymous
Gaul), embroidered with silver and gold. "By the crown of my empire,"
said Otto (according to the Gaul), "that which I see far exceeds what
rumor had reported." He took the imperial crown from his own head
and placed it on Boleslav's, gave him a nail from the Cross and the lance
of Saint Maurice, and established a Polish archbishopric. "In return,
Boleslav gave him the arm of Saint Adalbert."

From Poland Otto marched to Quedlinburg to celebrate Easter with
his sister, the new abbess. By the end of April he was in Aachen, where
he performed the astonishing act that symbolizes his short reign: He
opened Charlemagne's tomb.

Four sources before 1050 tell the tale. Some sound horrified, some
are matter-of-fact, others seem in awe. The *Annals of Hildesheim*, written after Otto's death (and not by his beloved teacher Bernward), condemned the act as "against holy religion and destined to bring upon its
doer vengeance for ever more."

Thietmar of Merseburg, who often traveled with the emperor, wrote
simply: "As he had doubts regarding the location of the bones of Emperor Charles, he secretly had the pavement over their supposed resting

place ripped up." He uncovered Charlemagne, seated on a throne. "After taking a gold cross which hung around the emperor's neck and part of his clothing, which remained uncorrupted, he replaced everything with great veneration."

A monk recorded the words of an eyewitness, the count of Pavia: "So we went into Charles. He did not lie, as the dead otherwise do, but sat as if he were living." His fingernails had penetrated through his gloves. "A strong smell struck us. We immediately gave Emperor Charles our kneeling homage, and Emperor Otto robed him on the spot with white garments, cut his nails, and put in order the damage that had been done. Emperor Charles had not lost any of his members to decay, excepting only the tip of his nose. Emperor Otto replaced this with gold, took a tooth from Charles's mouth, walled up the entrance to the chamber, and withdrew again."

Ademar of Chabannes told the fullest version. A dream told Otto to find Charlemagne, whose tomb had been lost. After fasting three days, Otto knew where to look. They found the emperor "sitting in a golden throne . . . crowned with a crown of gold and gems, holding a scepter and sword of purest gold, the body itself uncorrupted." They took his corpse from the crypt. "A canon of that church . . . who was enormous . . . compared his leg to that of the king, and his was found to be smaller. Immediately afterward, by a divine miracle, his leg was fractured." Anxious not to anger the sacred king further, Otto reburied Charlemagne behind the altar in "a magnificent golden crypt." It "began to be known by means of many signs and miracles," like the tomb of a saint.

Otto's odd and apparently sacrilegious act could be explained, in fact, as the first step toward canonizing Charlemagne as a saint. Or perhaps Otto thought he was, like so many heroes of story, descending to the Underworld to ask advice of the illustrious dead. We don't know what Gerbert thought about it; he never mentions this episode so apparently against all reason and logic. Yet for a Saxon-Greek who knew his Virgil and who believed that the world would not end so long as "the kings of the Franks who hold the empire by right shall last," it was not so totally illogical to acquire the blessing of the greatest of those kings.

It didn't help. Otto had not won the hearts of the Romans. They did not agree that Otto held "the empire by right," or that Charlemagne's heir shielded them from the Apocalypse. They did not like his choice of Gerbert as pope. As soon as Otto and his German army went north, they rebelled.

In June 1000, Gerbert called for help. Unfortunately, he chose a messenger who had already joined the other side. "Many facts that you heard garbled through lying rumor, I have entrusted to Gregory of Tusculum," he wrote Otto. "But I protest that what happened to us at Orte during the sacred solemnities of the Mass should not be accepted lightly." Unnamed persons had incited a riot in the church while Gerbert was presiding. "Within the holy of holies swords were drawn, and we withdrew from the city amidst the swords of frenzied enemies."

Otto raced back to calm Rome, but revolt soon broke out in nearby Tivoli. Otto surrounded the city, built earthworks and siege engines, and interrupted the water supply—again Gerbert supposedly put his technical skill to work. At the last minute, Otto allowed Gerbert and Bishop Bernward to mediate. They were "joyfully" welcomed into the city, says *The Life of Saint Bernward*. The next day they returned to Otto's camp in "a noble triumphal procession. All the leading citizens of the city followed them, naked but for loincloths, carrying in their right hands a sword and in their left a rod." They surrendered to the emperor, "excepting nothing, not even their lives." Otto praised the peacemakers, Gerbert and Bernward, "and at their request pardoned the guilty."

The ceremony was staged. Like all medieval acts of submission, nothing was left to chance. Gerbert and Bernward had guaranteed the emperor's forgiveness. The public ritual of surrender, "excepting nothing, not even their lives," was meant to restore the emperor's honor. Everyone involved knew the emperor would not refuse his pardon.

None knew this better than the Romans, who took it as a sign of weakness and rebelled in their turn. Their leader, Crescentius III, the son of the beheaded Crescentius of the Marble Horse, was backed by that same Gregory of Tusculum to whom Gerbert had entrusted his earlier news of rioting in the church at Orte. They trapped Gerbert in the Lateran and chased Otto from his palace to the Castel Sant'Angelo. Otto climbed to the tower and leaned out over the mob, crying: "Are you not my Romans? For your sake I left my homeland and my kinsmen, for love of you I have rejected my Saxons and all Germans. . . . I have adopted you as sons, I have preferred you to all others. . . . And in return now you have cast off your father and have cruelly murdered my friends. You have closed me out!"

His speech moved the mob to tears, according to *The Life of Saint Bernward.* The uprising was calmed, the rioters dispersed, and with the help of Duke Henry of Bavaria, Otto and Gerbert escaped to Ravenna.

Otto holed up in the monastery of Sant'Apollinaire in Classe, surrounded by its pine forests and salt marshes. His power was waning, and he had become desperate. He came under the sway of Romuald, another ancient ascetic revered as a living saint. Romuald had spent many years at Cuxa in Catalonia, where Gerbert's friend Garin was abbot; the monks there had prayed every moment that this super-holy man would die, so that his relics (his bones) would enrich their monastery's fame. But Romuald lived on and on, and finally he returned to Ravenna.

There he obliged Otto's pleas for more and more penance. Otto fasted. He chanted psalms. He wore a hairshirt under his imperial purple. Beneath his luxurious bed-coverings was a mat of hard reeds. Romuald advised him to renounce the world and become a hermit. "I will do that," Otto promised—after he had reconquered Rome. "From this hour I promise to God and his saints: After three years, within which I will correct the mistakes of my imperial rule, I will give up the realm to someone better than I am." He would give away all his money and "follow Christ in destitution, and with all my soul."

"If you go to Rome," said Romuald, "you will never return to Ravenna." It was the one demand Otto could not heed.

Otto marched on Rome. But Duke Henry had gone back to Bavaria, and Otto's own army was too small to break down the city gates. Terrible storms disrupted the siege, and Otto returned north. He spent Christmas at Todi, then marched again on Rome. He made it as far as the castle of Paterno, 40 miles from the city. Struck down by a violent fever—probably malaria—he confessed his sins, received absolution, and died in the arms of his friend, Herbert of Koln, who had just arrived, too late, with the full German army. The Wonder of the World was dead at twenty-two. With him died the dream of all three Ottos: to renew the Roman Empire. With him died all Gerbert's hopes.

CHAPTER XIII

The End of the World

The Christian empire that Gerbert imagined could have changed the course of history. As in the Spain he had seen in his youth, men of any faith would have risen to power. Trade in goods and ideas would have flourished. Justice and law would have curbed brutality and force. Instead of the Crusades, the Schism between East and West, and the Inquisition, the medieval Church would be known today for its arithmetic, geometry, astronomy, and music. God would still be worshipped through number, measure, and weight. Science and faith would be one.

He had been so close to success. Gerbert stayed behind in Ravenna when Otto marched on Rome. He was not at Otto's side when he died. Soon after the terrible news reached him, he heard that Otto's bride, Princess Zoe of Constantinople, had arrived in Italy. Through her, Gerbert's dream of Empire could have come true, for in later life she was heir to Byzantium. Her son and Otto's could have claimed—and fused—both empires, East and West, and recreated the Golden Age of the Caesars. But Otto, the Wonder of the World, was dead, and Zoe was sent home.

Meanwhile, Herbert's German army snuck away north—legend says the Germans tied Otto's corpse to his warhorse so the Italians would not realize the emperor was dead. Duke Henry of Bavaria, son of Henry the Quarreler, was acclaimed king of Germany through the connivance of Archbishop Willigis of Mainz, though Pope Gregory's father was heir to the throne. The rebellious Margrave Arduin of Ivrea became king of Italy. The Holy Roman Empire was shattered.

So, we can guess, was Gerbert's happiness. Grieving, impotent, his hopes gone, he would outlive his protégé by little more than a year.

Rome fell into the hands of Crescentius III, son of the beheaded Crescentius of the Marble Horse, and of Gregory of Tusculum, whom Gerbert had once thought a friend. The year before, Gerbert had bestowed a large benefice on this Gregory's son-in-law. Perhaps for that reason, the pope was not strangled, starved, imprisoned, exiled, or otherwise harassed by the Romans, as so many of his predecessors had been. Neither, however, was he consulted by the Crescentians or by the new king of Italy. His connections with the German king were cut off. Gerbert returned to the Lateran palace, with its high Romanesque towers from which he could study the stars, and resumed the work of Pope Sylvester II. He wrote charters, tried to settle church disputes, and received envoys, including the son of his first patron, Count Borrell of Barcelona. What gifts did the count's son bring from Spain? A book on mathematics or astrology? An astrolabe made in Baghdad? We do not know. None of Gerbert's personal letters from his final year remain.

His enemies later claimed that the pope spent his time practicing black magic. William of Malmesbury, writing in the 1120s, gives us one such picture of Gerbert's last days. Skewed, symbolic, it was immensely popular: Most later writers reference it. According to William, Sylvester II occupied himself in Rome practicing "the black arts" to enlarge his already substantial power and wealth: "He kept up the pressure for his own advancement with the devil's assistance, so that no project which he had conceived was ever left unfinished. For instance, his skill in necromancy enabled him to clear away the rubble and discover treasures buried by pagans long ago, which he used to satisfy his greed."

In the Fields of Mars near the Pantheon was a "statue," William says. It was "pointing with the forefinger of the right hand, and on its head were the words 'Strike here.'" It was all battered with blows from men who had done the obvious. Gerbert found "quite another answer to the riddle," William writes. "At midday with the sun high overhead, he observed the spot reached by the shadow of the pointing finger, and marked it with a stake." In the Fields of Mars is, in fact, an obelisk covered with hieroglyphics. It was placed there ten years before Christ's birth by Caesar Augustus. Gerbert, walking past one day, may indeed have perceived that it was more than a giant sundial.

The pope, says William, returned at night with a servant and, presumably, a shovel. They quickly found themselves in "a vast palace, gold walls, gold ceilings, everything gold; gold knights seemed to be passing the time with golden dice, and a king and queen, all of the precious metal, sitting at dinner with their meat before them and servants in attendance; the dishes of great weight and price." The palace was magically lit by a sparkling jewel; a golden boy stood opposite it, "holding a bow at full stretch with an arrow at the ready." Gerbert's servant, overcome with greed, snatched a golden knife. At once the figures came alive with a roar. The boy loosed his arrow and put out the light. And "had not the servant, at a warning word from his master, instantly thrown back the knife, they would both have paid a grievous penalty." They covered their tracks and said no more about it.

If this story were set in Reims, it could be dismissed as utter fantasy. But Rome is built on layers upon layers of palaces and treasure halls. In 2005, archaeologists were using a coring drill to survey the foundations of Caesar Augustus's palace on the Palatine Hill. Fifty feet down, the drill plunged into a void. Sending down a camera, the crew discovered a sacred grotto—a round, domed room about twenty-five feet high and twenty-five feet across, covered with mosaics of marble and seashell. In the soft light of the remote-sensing probe, they glittered like gold. Perhaps it was something like this that the obelisk had revealed to Gerbert.

It may also be that William of Malmesbury was making a very complicated literary allusion. Given that he was writing for educated monks, his pointing finger should have summoned up a lesson from Saint Augustine: "To those who do not understand what I am writing, I say this:

It is hardly my fault if they do not understand. It is as if they wish to see the old or new moon or some faintly shining star that I am pointing at with my outstretched finger, but find their sharpness of vision insufficient even to see the finger itself."

Gerbert, like Saint Augustine, moved beyond the literal meaning of "Strike here" to the figurative: striking, not the finger itself, but what the finger pointed to. Yet, because his motives were not pure, Gerbert received no benefit from his superior wisdom. In William's metaphor, the gold Gerbert saw in the vault beneath the Fields of Mars stands for the scientific treasures that Gerbert had "dug up" by studying classical and Arabic texts. But God did not allow Gerbert and his servant to take even one golden knife from that vault. Gerbert's sin was not the study of science (entering the vault) but his *reason* for studying it. Gerbert was concerned only with "his own advancement," says William. He did not heed Augustine's teaching that knowledge should always be subservient to the Word of God. And so he was punished.

Gerbert was so blinded by hubris and greed that, for all his learning, he saw not Truth. Such is the moral of a second of William's stories. "After close inspection of the heavenly bodies," he writes, Gerbert cast a metal head "which could speak, though only if spoken to." Asked a yes or no question, it "would utter the truth": Will I be pope? Yes. Will I die before I sing Mass in Jerusalem? No. "This answer, they say, was ambiguous and misled him," William writes. "He gave no thought to repentance, flattering himself with the prospect of a long life; for why should he go deliberately to Jerusalem to hasten his own death? He did not foresee that there is in Rome a church called Jerusalem."

So it was that on May 3, 1003, a little over a year after Otto III's death, Gerbert fell ill while saying Mass at the Church of the Holy Cross of Jerusalem in Rome. This much is true.

William elaborates: "Sending for the cardinals, he lamented at length his own misdeeds." Then he gave orders "that he should be cut in small pieces and cast out limb by limb. 'Let him have the service of my body,' he cried, 'who sought its obedience; my mind never accepted that oath, not sacrament but rather sacrilege.'"

The Fourth Trumpet of the Apocalypse, from an eleventh-century manuscript made at the monastery of Saint-Sever, France. As Pope Sergius wrote in Gerbert's epitaph, upon his death "the world was darkened and peace disappeared."

Gerbert died May 12 and was buried, at his request, in the Lateran, the church built by Emperor Constantine for the first Pope Sylvester. His tomb was destroyed when the church was enlarged in 1648. By then William of Malmesbury's story of his death was so widely believed that a canon who saw the tomb opened was careful to report that

Gerbert's body had *not*, in fact, been cut up into little pieces in penance for his unholy acts of magic. The body lay in the coffin whole, he wrote, dressed in papal raiment, with ring and staff. Exposed to air, it disintegrated into dust, leaving only a sweet perfume.

The marble tombstone, saved during the reconstruction of the church, now hangs on a pillar in the right aisle. Pope Sergius IV, who had been Gerbert's papal librarian, wrote the epitaph. Through his words, we see how Gerbert's peers—the churchmen who would have been at his bedside when he died—actually saw him.

Sergius makes no mention of black arts, sacrilege, hubris, greed, or Gerbert's body being cut into bits. He wrote: "The wise Virgin and Rome, the head of the world, made him celebrated throughout the universe. . . . The emperor, Otto III, to whom he was always faithful and devoted, loved him greatly and offered him this church of Rome. They illuminated their time, emperor and pope, by the brilliance of their wisdom. The century rejoiced." Upon Gerbert's death, Sergius said, writing six years later, "the world was darkened and peace disappeared."

At Gerbert's death, a world did end—the world in which Muslims, Christians, and Jews could sit down together and translate works of science from Arabic and Greek into Latin; the world in which a peasant boy who had excelled in such science could end up as pope.

In just thirty years, the idea that the best candidate for pope could be a scientist or philosopher would be unthinkable. The new worldview is hinted at in a famous poetic line from Ralph the Bald's history. The earth was so relieved that the Apocalypse had not come in A.D. 1000, a thousand years after Christ's birth, said Ralph, that it "put on a white mantle of churches." And many churches were built or expanded around the year 1000, as we have seen. By 1036, when Ralph wrote this line, churches, new or old, provided the sole source of authority and comfort to the common man.

At the start of the eleventh century, the kings and counts of Germany and Italy fought for the right to succeed Otto III as emperor. The emperor and pope—or antipope—battled for precedence. Famine, drought, and flood besieged Europe. St. Anthony's Fire (the painful disease of ergotism, caused by rotten rye) reappeared. Comets, earthquakes, eclipses, unusually large whales, wolves in churches, and rains of blood and stones were taken as signs and wonders prophesying the Apocalypse still to come.

Churches became less famed for their libraries and schoolmasters, and more for their holy relics alone. Chartres, led by Gerbert's student Fulbert from 1006 to 1028, became the center for the cult of the Virgin (it owned the tunic she wore at the Annunciation). In 1016, a little village in Aquitaine suddenly "found" the head of John the Baptist in a cellar. Though Antioch already displayed one of John's heads, this skull was quickly housed in a jewel-covered reliquary and placed on the altar. Ademar of Chabannes, who praised Gerbert as the "scientist pope," forged ancient documents and composed a glorious processional Mass in order to turn his church's founder, Saint Martial, from mere saint into an apostle and companion of Jesus. In 1018, a crowd waiting to touch the blessed Martial's tomb turned into a mob. Pilgrims rushed the doors "like a river flowing into the church," Ademar wrote, "by accident falling over itself, each person trampled the other." More than fifty died.

Saints' relics not only enriched their churches and excited the masses, they brought the Peace of God to warring factions. Monks carried them on poles at the heads of parades; huge crowds gathered at these "relic jamborees," where counts and castellans were forced to swear to keep the saints' peace. "Ye masses of commoners, give thanks to God," begins the peace hymn of Fulbert of Chartres. "He has summoned you to his aid; he puts on you the burden of peace and order. The nobles, long unaccustomed to obeying the law, have made good resolutions and will keep them. Thieves submit . . . travelers go about untroubled . . . the vine flourishes. . . . The lance is turned into a scythe, the sword into a ploughshare. This peace enriches the poor and impoverishes the robbers." Miracles occurred at the gatherings: The blind saw, the crippled walked. The meetings grew larger and rowdier and the churches more powerful as *Peace!* became a call to arms. In 1026, seven hundred French knights were sent on pilgrimage to Jerusalem in the name of peace; the next time they would be called crusaders. Even the Mass was revised: Instead of reciting *Miserere nobis* ("Have mercy on us") the third time, the monks now sang *Dona nobis pacem* ("Give us peace").

The Church Gerbert grew up in was gone. Churchmen who opposed this new kind of Catholicism, who repudiated the new rituals of relic worship, infant baptism, sanctification of marriage, intercession for the dead, confession to priests, and veneration of the crucifix; who saw Christ as the all-powerful King of Heaven, not the broken human sacrifice on the cross; who thought monks should demonstrate their faith, not by managing great estates and acquiring riches, but through prayer, study, and teaching—these churchmen were denounced as heretics.

Accusations of heresy soon became the best way for the Church to make its intelligentsia toe the line. The first heretics ever burned at the stake were a party of noblemen and educated clerics—among them two of Gerbert's students—in Orleans in 1022. They were condemned by another of Gerbert's students, King Robert the Pious of France. One of the burned was the queen's confessor. These burnings of church intellectuals would lead to the founding of the Inquisition in the 1200s. In 1210 the reading of Aristotle, or even commentaries on Aristotle such as Boethius—the main textbook in the cathedral schools during the previous five hundred years—was made punishable by excommunication. The enlightened Dark Ages of Gerbert's time had been replaced by a much darker era, during which scientists at the newly founded universities fought with the Church to reintroduce the astronomy and mathematics that Pope Sylvester II had known.

Fear of the Apocalypse—and the corresponding hope to be among the Chosen taken by Jesus into Heaven—was not the only force behind this new religious intolerance. When Gerbert and Otto III succeeded in bringing Hungary into the Christian world, they opened a land route to Jerusalem. Walking in Jesus' footsteps soon became so popular that the Muslim ruler of Jerusalem grew annoyed by the constant stream of Christian tourists. To reduce his city's draw (and, say some sources, to bedevil his Christian mother), in 1009 he destroyed the Holy Sepulchre, where Jesus had been buried after his crucifixion.

News of this outrage reached France with a strange twist: It was blamed on the Jews, rather than the Muslim emir. In Limoges, Orleans, Rouen, and Mainz, Jewish citizens were attacked and forced to convert, or expelled and their property confiscated. Before the year 1000, there were no Jewish quarters in medieval cities: Jews lived wherever they wished. Jewish landowners were not uncommon. They owned Christian serfs and owed military service to their Christian overlords. (Remember the Jewish knight who gave up his horse so that Emperor Otto II could escape the Saracens in 982?) There were Jewish artisans, merchants, and, especially, doctors. Not until after the First Crusade in 1096 were Jews forbidden to carry weapons and restricted to the sinful trade of moneylending. At Rouen, the crusaders "herded the Jews into a certain place of worship . . . and without distinction of age or sex put them to the sword," wrote Guibert of Nogent late in the eleventh century. At Worms the crusaders killed eight hundred Jews; additional massacres at the end of the century were recorded at Mainz, Koln, Trier, Metz, Bamberg, Regensburg, Prague, and many other towns.

A similar act of intolerance destroyed the peace between Christian and Muslim Spain. The regent al-Mansur the Victorious, who had seized power in Cordoba in 976 and sacked Barcelona in 985, escalated the border wars in 997 by destroying the shrine of Santiago of Compostela. Fighting against Christians—particularly since he frequently won—was his way to stay in power. Upon his death in 1002, his son overstepped, claiming the title of "caliph," not "regent." Al-Mansur's army of mercenary Berbers saw that as blasphemy—a caliph had to be descended from Muhammad or one of his family members—and rebelled, destroying not only the new regent, but the fine city of Cordoba itself. The library of 40,000 (or 400,000) books was burned. The years of *conviviencia*, or tolerance among Muslim, Christian, and Jew, were over; it was time for the *reconquista*, the Reconquest. In 1063, on their way to fight the Muslims in Spain, Christian knights from France took the opportunity to destroy every Jewish household they found along their way—an infidel was an infidel, after all.

Finally, all hope that a future Roman emperor could take a Byzantine bride and so reunite East and West, fusing all of Christendom into one cosmic empire, was dashed by three little words known as the *filioque*—"and the Son"—added to the Roman creed in 1020. In Constantinople, the Holy Spirit came from the Father *to* the Son; in Rome, it came from the Father *and* the Son, meaning that the three members of the Holy Trinity were not ranked, one-two-three, but coequal. This tiny change in wording led to a thirty-year argument between the two churches. No compromise could be reached, and a formal schism between the Roman Catholic and Eastern Orthodox churches was announced in 1054. It lasts to this day.

As the medieval world changed, Gerbert's reputation changed as well. Writing during Gerbert's lifetime, the monk Richer of Saint-Remy described him as "a man of such great genius and admirable eloquence that his glory blazed over all of Gaul like a burning flame." Richer may have sought to flatter Gerbert, but the chronicler from Fleury, where Gerbert's enemy, Abbo, was still abbot, had no need to. Yet, announcing his election as pope, he wrote, "Otto, inspired by the reverence without bound that he professed for Gerbert, on account of his incomparable scientific knowledge, elevated him to the papacy." Upon Gerbert's death, said Pope Sergius IV, "the world was darkened," while in Catalonia, the son of Count Borrell of Barcelona mourned the "glorious and very wise pope Gerbert, who was called Sylvester."

Thietmar of Merseburg, writing between 1013 and 1018, remarked that Gerbert was a skillful astronomer and "surpassed his contemporaries in his knowledge of various arts." Between 1030 and 1046, Ralph the Bald—the mystic who recorded so many signs and wonders of the Apocalypse—wrote simply, "This Gerbert came from a rather undistinguished family in Gaul, but he was acutely intelligent and deeply learned in the study of the liberal arts." Between 1070 and 1100, Abbot

Sigebert of Gembloux would write a chronicle calling the turn of the millennium a Golden Age of Science. "Happy the state whose rulers are philosophers, and whose philosophers rule," he said. His long list of churchmen famous for their knowledge ends with Gerbert, "who among those shining, shone exceedingly."

At about the same time, however, Cardinal Beno wrote his diatribe *The Deeds of the Church of Rome vs. Hildebrand.* Hildebrand was the given name of the sitting pope, Gregory VII, who was in the process of making the papacy the supreme power in Europe. Emperor Henry IV wanted to replace him with the more tractable antipope, Clement III. To prove Hildebrand's unfitness, Beno claimed he had been schooled by the disciples of Gerbert, celebrated for his wizardry. This Gerbert, wrote Beno, had made a pact with the devil. For what else could explain his "dazzling ascent from lowly beginnings" and the magical tricks he used to tell time by the stars? Had not Gerbert the Wizard once joked that his career had progressed from R to R to R—Reims to Ravenna to Rome? What could this be but a demonic spell? (Miro Bonfill would have thought it a charming pun; Adalbold would have made a triangle out of it.)

Beno was also the source of the pernicious story that Gerbert could not die, so the devil promised, before he had sung Mass in Jerusalem. Wrote Beno, "Forgetful that the Church of San Croce was known as *in Gerusalemme,* he said Mass in it. Immediately after, he died a most horrible death, ordering with his last breath his hands and tongue, with which by sacrificing to demons he had dishonored God, to be cut to pieces."

Tales of Gerbert's devil grew in the telling. Walter Map, writing in the twelfth century, personified her as Meridiana, "a woman of un-heard-of beauty seated on a large silken carpet, and having before her a huge heap of money." The "notorious Gerbert," Walter wrote, was "busily engaged at Reims in the effort to surpass in intellect and utter-ance all the students of the school, whether native or foreign, and was successful," when he made his sinful bargain. "Thereafter, free and af-fluent in Meridiana's gifts, he enriched himself with household goods, and a crowd of servants, gathered to him changes of raiment and money, and grew strong with food and drink: so that his wealth in Reims was like the glory of Solomon in Jerusalem, and his settled joy in love not inferior." (To be fair to Walter, this passage echoes the complaints of the nobles of Bobbio that Gerbert had brought from Reims such an extravagant household that they thought he must be keeping a wife.) Continues Walter: "Every night she, who possessed full knowledge of the past, instructed him in what he was to do by day. . . . Within a short time no one was his equal."

It was Meridiana, Walter says, who promised Gerbert he could not die until he sang Mass in Jerusalem. When he was struck down at the altar of the church in Rome, Gerbert saw her exultant face in the crowd. Walter omits the request for mutilation and gives Gerbert a good death: He "sincerely hallowed the short remainder of his life with assiduous and severe penance, and died in a good confession." He concludes: "Though through covetousness Gerbert was held captive a long time by the birdlime of the devil, he ruled the Roman Church greatly and with a strong hand." But his marble tomb in the Lateran, Walter added, "continually sweats."

William of Malmesbury went even further than Walter Map, turning Gerbert's years of learning mathematics in Spain into a series of magical escapades with beautiful women, Saracen wizards, and secret codes hidden in the constellations. Gerbert, in this version, lodged in the house of a Saracen philosopher who "sold his knowledge" and lent Gerbert books to copy. "There was, however, one volume to which he had committed all his art and which Gerbert could by no means get out of him. On his side he was passionately anxious to make the book somehow serve his turn. 'We ever strive toward what is forbidden.'" (It is probably true that Gerbert never saw a science book he didn't covet.) With the help of the Saracen's beautiful daughter, Gerbert stole the book. The Saracen used "the stars" to track him, but Gerbert hung by his hands under a bridge, knowing that the stars would say he was neither on land nor water. He reached the coast, summoned the devil, and had himself and the magical book safely conveyed "overseas."

Then there is the magical talking head William says Gerbert made in Rome "after close inspection of the heavenly bodies." Strangely, a very similar talking statue that answered yes-or-no questions appears in that "wonderful Roman book of history" by Orosius, which Bishop Race-mundo brought back to Cordoba in 949 as a gift from the emperor of Constantinople. Ibn Juljul mentions it; he was the one who told how the Jew Hasdai arranged for Dioscorides' *On Medicine* to be translated from Greek to Arabic. "The pagans of Rome," wrote Ibn Juljul, "pre-tended that the statue was erected according to the movements of the stars and that it contained the spirit of one of the seven planets." How such a thing got linked to Gerbert, we will never know; perhaps he had heard the tale in Spain and told it himself. Or Gerbert's talking head might have been mixed up with the majesty of Count Gerald the Good in Aurillac, "an image made with such precision to the face of the human form that it seemed to see with its attentive, observant gaze."

Regardless, the magical statue, the pact with the devil, and the grue-some death in "Jerusalem" proved more potent than the abacus, Arabic numerals, experimental geometry, the celestial spheres, or the astrolabe in making Gerbert's name. From the thirteenth century on, Gerbert would be the Magician Pope, Gerbert the Wizard, "the best necro-mancer in France, whom the demons of the air readily obeyed in all that he required of them by day and night."

Martin the Pole, a Dominican friar and the pope's confessor in the thirteenth century, codified what has come to be known as the Dark Legend of Gerbert. Martin wrote a *Lives of the Popes* that lasted for hun-dreds of years as the official version. A copy made in the mid-1400s says of Gerbert, "He quitted the monastery and made homage to the devil so that he could obtain all his desires." According to Martin, Ger-bert studied, not in Cordoba, but in Seville, "where he made such progress that his wisdom impressed even the highest dignitaries." He taught Emperor Otto III and King Robert of France—but all of these great honors Gerbert received only through the wiles of the devil, who promised him that he would not die before he sang Mass in Jerusalem. Gerbert, writes Martin, was pleased: He had not the slightest intention of taking a pilgrimage. Then, celebrating Mass in the church called "Jerusalem," he heard "the strident cries of the devils":

> He began to groan and sigh. But even though he had been the worst of sinners, he did not despair of the mercy of God. Confessing his sins in public, he ordered that they cut off his limbs, with which he had made homage to the devil, and that they place his dead trunk in a cart. They should bury him wherever the beasts who hauled the cart stopped. This was done. He was buried in the Lateran, as a sign that he had received God's mercy. By the sound of knocking bones and the sweating of the marble, his tomb presages the death of any pope.

Martin brought all the elements together: Gerbert's hubris and sin, the exulting devils, the confession, the mutilation, the mercy of God, and the sweating tomb—to which he adds the knocking bones. In these accounts, "Pride goeth before a fall" is the sole moral of Pope Sylvester II's life.

The Dark Legend survived unchallenged until 1570, when one of Luther's disciples discovered *The Acts of Saint-Basle* in a German archive. Bishop Arnulf of Orlean's diatribe against the popes—"Deplorable Rome! . . . What spectacles have we not witnessed in our days!"— became a weapon in the war between Protestants and Catholics, and Gerbert was named its author. Suddenly his life was seen in a new light. That "pact with the devil" was nothing but a Catholic slur on a great man of science who saw through the rigamarole and hypocrisy of the medieval Church.

Protestant intellectuals went looking for more of this kind of ammunition against Rome. Soon Constantine's copy of Gerbert's letter collection was rediscovered in a private library—no mention of Gerbert's letters had been made in any text since the twelfth century. In 1587, twelve of the most stridently anti-papal ones were printed by Nicolas Vignier, a staunch Calvinist who was physician and historian to the king of France. One of the letters Vignier chose was Gerbert's attack on Peter of Pavia (soon to become Pope John XIV): "You do not cease from thefts from our church. . . . Steal, pillage, arouse the forces of Italy against us!" Another decries the "deceit and changeableness" of the church in Italy and ends, "Bewail the future ruin not so much of buildings as of souls; and do not despair of God's mercy."

At the same time, another set of Gerbert's letters surfaced in France—containing some not found in Constantine's collection; it is thought that Gerbert had originally made it for a friend in Germany. This manuscript was lent to Cardinal Baronius, the Vatican librarian. Baronius made the copy that now resides in the Vatican. He returned it to France, where the letters were printed in an edition of 1611; the original manuscript disappeared soon afterward.

Baronius completed his immense *Annals of the Church* in 1602—a few years before Galileo invented the telescope. The *Annals* introduced the Terrors of the Year 1000 to the modern world—ironically, by pooh-poohing them. The idea that the world would end in A.D. 1000, Baronius said, was "first announced in Gaul, first preached about in Paris, then proclaimed by the vulgar throughout the world, believed by many, certainly accepted by the simpler minded people in fear, but it seemed improbable to the learned." He cites Abbo of Fleury's letter to the king, claiming to have heard in his youth a preacher in Paris announce that the End of the World would occur at the year 1000. He cites Ralph the Bald, whose signs and wonders included rains of blood, wolves in churches, and unusually large whales. Ralph's work had been ignored for five centuries; Baronius gave it new life.

Of the Magician Pope, the Vatican librarian had the same opinion as the Protestants: "Gerbert was nothing but a learned man who was ahead of his time. Those who want to efface his name from the catalogue of popes are ignorant fools perverted by the schismatic Beno. Who can give credence to that fable of the Fields of Mars, of the servant and the pretty knife? There is between this story and that of the lady pope, Joan, one difference: that is, the latter one was imagined to give joy, the other to denigrate."

Another Catholic scholar, Geraud Vigier, added a note about Gerbert to his history of Saint Gerald of Aurillac in 1635. Those old stories of magic and the devil, he wrote, reflected "the beliefs of a simple people in a barbaric century, who believed that anyone who could, following the rules of arithmetic, count how many tiles there were on a roof, was a magician."

With Catholic and Protestant scholars in agreement, the Magician Pope should have been forgotten and only the Scientist Pope bequeathed to posterity. But by putting the Dark Legend into print, Baronius and Vigier unintentionally lengthened its life. Like the idea of the Terrors of the Year 1000, the story of Pope Gerbert the Wizard was too good to rationalize away.

Jules Michelet, in his *History of France* of 1831, increased the appeal of both by linking them. Michelet created an indelible picture of the Year 1000: "This horrifying hope of the Last Judgment rose throughout the calamities that preceded the Year 1000 or followed right after it. It seemed that the order of the seasons had been overturned, that the elements followed new laws. A terrible plague devastated Aquitaine, the flesh of the diseased burned with fire, came away from their bones, and rotted away." Famine ravaged all the world, followed by cannibalism. Peasants fled their farms, donating their lands and goods to the Church in hopes of being spared.

If there had been a real pope in Saint Peter's chair in 999, Michelet implied, people's fears might have been calmed. But "this Gerbert," Michelet said, "was nothing less than a magician. A monk of Aurillac, he was expelled and found refuge in Barcelona, where he defrocked himself to study rhetoric and algebra in Cordoba." He taught Otto II and Otto III, established his famous school at Reims, and became pope. "One day, as he was celebrating mass at Rome in a chapel known as Jerusalem, the devil appeared and reclaimed the pope. It was a bargain that they had made in Spain, among the Muslims."

In spite of modern scholars' efforts to defeat it—including the publication of all of Gerbert's known works in Latin between 1867 and 1899, and the translation of his letters into English in 1961 and French in 1993—the Dark Legend persists. It appeals to the same mindset that sees science "at war with" religion, or that accepts Washington Irving's fiction about Columbus facing down a council of benighted church-men who were insisting the earth was flat.

As late as 1988, a writer could present as history a dramatic scene that would have made Irving proud. Citing "an ancient chronicle," he describes Saint Peter's on the eve of the Year 1000, "thronged with a mass of weeping and trembling worshipers." For many of them, Ger-bert, celebrating midnight Mass, "fulfilled the dark prophecy." He was the Antichrist.

> Did he not voice disbelief in the imminent end of the world? Had he not introduced evil wisdom gotten from Jews and Moors? Did he not practice black arts such as astronomy? It was rumored that, in Spain, Sylvester had taken a Mahometan witch for his mistress who had taught him to fly through the air and foresee the future. He was said to have made an artificial man who could speak and answer ques-tions put to him with a yes or a no. He was making spheres depict-ing the world as an orb. Surely this was blasphemy. It was thought that he had sold his soul to the devil in order to obtain knowledge and honor. And so the worshipers trembled in fear.

There is, unfortunately, no "ancient chronicle" in which Gerbert says Mass on New Year's Eve 999, in Saint Peter's or anywhere else. No one in Gerbert's lifetime saw him, seriously, as Antichrist (though it was a common insult). Gerbert never voiced "disbelief at the imminent end of the world"; he never mentioned it at all. No one thought Jews or Muslims or wisdom were evil. Astronomy was a central part of the cur-riculum at all cathedral schools. The witch and the artificial man were inventions of the late eleventh century. Everyone knew the world was an orb. No matter. The Magician Pope sells books.

As the year 2000 approached and scholars looked back to the first millennium, historians held conferences on "The Apocalyptic Year 1000," leaving it up to scientists to rediscover the Scientist Pope. The keynote speaker at the Thirteenth Annual microCAD International Computer Science Conference, held in Hungary in 1999, called Gerbert "the Y1K person most representative of the group gathered here today" and "the Bill Gates of the end of the first millennium." The Millennium Issue of the magazine *Searcher* gave him credit for developing Europe's "Computer Number One." In 2000, *Sky & Telescope* printed an article about "Y1K's Science Guy," which brought Gerbert's story to the attention of an astrophysicist in Rome. With the assistance of the Vatican, he subsequently organized a series of lectures and events to commemorate the millennium of Gerbert's pontificate, including a grand Requiem Mass held in the cathedral of Saint John Lateran in 2003. A 1999 letter from Pope John Paul II summed up the official church position: Gerbert was "remarkable" for "the breadth of his knowledge." He was "a learned humanist and wise philosopher, a true promoter of culture, [who] put his intelligence at the service of the human person. . . . He reminds us that intelligence is a marvellous gift from the Creator."

Gerbert made news again after 9/11. With stricken Americans tending to see all Muslims as terrorists, the Mathematical Association of America's newsletter took the occasion to remind its readers that "as mathematicians, we are all children of Islam." Gerbert's cathedral school, the essay continued, was proof that Islam played "a crucial role . . . in the development of the West's scientific tradition."

A professor at Nehru University in New Delhi told Gerbert's story once again in October 2006. Her editorial in the Pakistan *Daily Times* responded to Pope Benedict XVI's now-infamous speech in Regensburg, Germany, in which he quoted from a medieval text saying, "Show me just what Muhammad brought that was new, and there you will find things only evil and inhuman." The pope should have searched a little harder in the Vatican Library, wrote the professor, and he might have come upon the story of Gerbert of Aurillac and learned how Islam brought the West the science on which our modern lives depend.

By December 2008, Pope Benedict had learned of Gerbert. In a message on the World Year of Astronomy, 2009, he mentioned that "among my predecessors of venerable memory there were some who studied this science, such as Sylvester II. . . . If the heavens, according to the Psalmist's beautiful words, 'are telling the glory of God' . . . the laws of nature which over the course of centuries many men and women of science have enabled us to understand better are a great incentive to contemplate the works of the Lord with gratitude."

Still the Dark Legend remains. In 1843, annoyed by the portrayal of their townsman in Michelet's *History of France*, the citizens of Aurillac raised funds to have a statue erected in the town square. In 1851, a crowned and dignified twelve-foot-tall bronze pope, sculpted by the artist David d'Angers, was set on top of a granite plinth. His hand is not raised in blessing, but gesturing, as if lecturing on a scientific point. Three bas-reliefs on the faces of the plinth portray him as the shepherd lad discovered by the monks of Saint-Gerald's while studying the stars with an anachronistic telescope; as the inventor presenting his pipe organ, celestial sphere, and (anachronistic) mechanical clock to the emperor; and as the pope carried in grand procession through the crowded streets of Rome.

Yet in 2005, a small community on the outskirts of Aurillac commissioned Hungarian artist Szinte Gabor to cover the walls and ceiling of their church with frescoes depicting scenes from Gerbert's life. Gabor drew the shepherd boy. He drew the pope crowning King Stephen, the first Christian ruler of Hungary. And he drew the studious young monk in Spain, a manuscript in his lap, scientific instruments scattered all about, a chart of Arabic numerals leaning up against his lectern, and a squat little devil at his feet.

ACKNOWLEDGMENTS

I was introduced to Gerbert of Aurillac through an act of grace. Writing a previous book, about an adventurous Viking woman, I found myself making an imaginary pilgrimage to Rome just after the year 1000. Wondering which pope (if any) Gudrid the Far-Traveler had met, I discovered the Scientist Pope. Gudrid, unfortunately, had arrived twenty years too late; in 1003 she was still in the New World. How might history be different if Gerbert had heard, firsthand, about the new land and people across the sea to the west?

In pursuing Gerbert's story, I am indebted to the many scholars, from Nicolai Bubnov in 1899 to Marco Zuccato, whose dissertation is unpublished as of this writing, who have worked to bring Gerbert's Dark Ages to light. I am especially grateful to Harriet Pratt Lattin for translating Gerbert's letters into English: I would not otherwise have heard his voice so clearly.

A special thanks to the people I interviewed as I traveled in Gerbert's footsteps and tried to understand his science: Sabina Bertozzi (Archivum Bobbiense series, Bobbio), Charles Burnett (Warburg Institute, London), Menso Folkerts (University of Munich), Paul Freedman (Yale University), David Juste (University of Sydney), David King (University of Frankfurt, emeritus), Paul Kunitzsch (University of Munich, emeritus), Christian Lauranson-Rosaz (University of Lyon), Jessica Lavelli (CoolTour, Bobbio), Maria Rosa Menocal (Yale University), Flavio Nuvolone (University of Fribourg), Romain Pageaud (Elne Cathedral), Julio Samso (University of Barcelona), and Costantino Sigismondi (University of Rome). Some went well beyond the role of "source," meeting my train, taking me to dinner, finding me a place to stay, introducing me to their colleagues, inviting me on a country hike or a tour of the city, sending me articles and books and photographs, commenting on rough drafts, and just checking in as my writing progressed. I would like to particularly single out Professor Nuvolone. He is said by his peers to be able to talk about Gerbert "as if he were still alive and in the next room," an observation I found to be true. I also give my best wishes to Professor Sigismondi, who hopes to see Gerbert sainted. Finally, though I was unable to meet him in person, I would like to thank Marco Zuccato, who did his best to keep me up-to-date with the latest scholarship through email. I consider all of these scholars—as well as the many others whose books and articles I consulted—to be my teachers. I could not have written this book without their contributions. I hope they will forgive me where my interpretation of Gerbert and his times does not match theirs, or where my book diverges from what they had expected.

Thanks also to Linda Wooster, Loralee Tester, Giovanna Peebles, Peter Travis and Monica Otter (both of Dartmouth University), Susan Bianconi (Yale University), Ginger McCleskey, William Fergus, James Collins, and Jeffery Mathison for their help with translations, travel, library access, and illustrations; to my editors, Lara Heimert, whose vision shaped the book, Brandon Proia, whose diligent editing brought it to completion, and Katherine Streckfus, who caught my mistakes; to my agent, Michelle Tessler, who guided me through the process of creation; and above all to my husband, Charles Fergus, who gave me confidence.

ILLUSTRATION CREDITS

Color Illustrations (center section)

Plate 1   *Majesty of Saint Foy.* Abbey Sainte Foy, Conques. Erich Lessing / Art Resource, New York.

Plate 2   *Dedicatory page from Saint Bernward's evangelary.* Hildesheim DS 18, fol. 16v. Dom Museum Hildesheim.

Plate 3   *Aristotle teaching on the astrolabe.* Ms. Ahmet III, 3206, Topkapi Palace Museum, Istanbul. Bridgeman-Giraudon / Art Resource, New York.

Plate 4   *Arqueta.* Girona cathedral. Akg-images / Bildarchiv Steffens.

Plate 5   *Gerbert's abacus board.* BnL Réserve précieuse Ms 770r. Bibliothèque nationale de Luxembourge.

Plate 6   *Model of the universe,* from *The Phenomenon of Aratus,* Ms. 188, fol. 30r, Bibliothèque municipale, Boulogne-sur-Mer. Giraudon / Art Resource, New York.

Plate 7   *Map of the world.* Preussicher Kulturbesitz Phill. 1833 (Rose 138), fol. 39v. Staatsbibliothek Berlin / Art Resource, New York.

Plate 8   *The coronation of Otto III.* Munich, Bayrischen Staatsbibliothek Cod. lat. 4453, fol. 24r. Bpk / Lutz Braun / Art Resource, New York.

Black and White Illustrations

viii   *Map by Jeffery Mathison.*

19   *The Church of Saint-Michael-of-the-Needle, Le Puy.* By the author.

34   *Saint Luke, from Saint Bernward's evangelary.* Hildesheim DS 18, fol. 118v. (detail). Dom Museum Hildesheim.

55   *Keyhole arch at Cuxa.* By the author.

62   *Graffiti at the cathedral of Elne.* By Romain Pageaud. Retouched by James Collins.

71   *Incipit to Gerbert's letter collection.* Leiden Voss, Q54, fol. 52v. Leiden University Special Collections.

85 *Arabic numerals from a mnemonic poem.* MS Trier 1093/1694 fol. 198r. Anja Runkel, Stadtbibliothek / Stadtarchiv, Trier.

107 *Page from Gerbert's geometry textbook.* University of Pennsylvania Lawrence J. Schoenberg Center for Electronic Text and Images manuscript #ljs 194, folio 10r.

110 *Gerbert's letter to Adalbold.* University of Pennsylvania Lawrence J. Schoenberg Center for Electronic Text and Images manuscript #ljs 194, folio 51v.

117 *Ptolemy.* Ulm Cathedral. Erich Lessing / Art Resource, New York.

141 *The Destombes astrolabe.* IMA AI 86–31. Institut du Monde Arabe, Paris.

164 *Gerbert's* Carmen Figuratum. Paris, BnF Lat. 776, f. 1v. Bibliothèque nationale de France.

167 *Arabic numerals hidden inside Gerbert's* Carmen Figuratum. By Flavio Nuvolone.

180 *Ivory carving of Otto II and Theophanu.* Réunion des Musées Nationaux / Art Resource, New York.

200 *Processional cross.* Cathedral Treasury, Palatine Chapel, Aachen. Erich Lessing / Art Resource, New York.

225 *The Last Judgment from the Bamberg Apocalypse.* Staatsbibliothek Bamberg. Snark / Art Resource, New York.

235 *The Fourth Trumpet.* Apocalypse of Saint-Sever. Bibliothèque nationale de France. Snark / Art Resource, New York.

NOTES

Introduction: The Dark Ages

1 *the world would end:* On the Terrors of the Year 1000, see Michael Frassetto, ed., *The Year 1000*, and Richard Landes, Andrew Gox, and David C. Van Meter, eds., *The Apocalyptic Year 1000*. Signs and wonders found throughout *The Five Books of the Histories* of Ralph the Bald, aka Rodulfus Glaber, were translated by John France, Neithard Bulst, and Paul Reynolds in a 1989 edition (hereafter cited as "Ralph the Bald"). Gerbert refers to the "falling away" of churches in his *Acts of Saint-Basle*, though he doesn't link it to the Apocalypse; see Pierre Riché, *Gerbert d'Aurillac*, 134. For Saint Augustine and Adso of Montier-en-der, see Benjamin Arnold, "Eschatological Imagination and the Program of Roman Imperial and Ecclesiastical Renewal at the End of the Tenth Century," in Landes et al., 247; for Adso's complete letter, see Adso of Montier-en-Der, "Book of the Antichrist," translated by Bernard McGinn, in *Apocalyptic Spirituality*, 89–96. Landes cites Abbo of Fleury and the annals of Anjou, 965 ("Fire from heaven . . . "), in "The Fear of an Apocalyptic Year 1000," in Landes et al., 250. "Greed . . . " is found in a 997 manuscript from Saint-Hilaire in Poitiers; see Richard Newhauser, "Avarice and the Apocalypse," in Landes et al., 113. For "Satan . . . " a comment by Ralph the Bald, see Henri Focillon, *The Year 1000*, 64. Jane Schulenburg traces "Clear signs . . . " in charters from 1025–1033 to a standard formula from the seventh century; see "Early Medieval Women, Prophecy, and Millennial Expectations," in Frassetto, 246.

3 *"geometrical figures":* Gerbert of Aurillac, *The Letters of Gerbert*, translated by Harriet Pratt Lattin, 299 (hereafter cited as "Gerbert"). I have used Lattin's translations, except where noted. Most of Gerbert's letters are undated; Lattin has attempted to date them more precisely than other editors, considering, as she says in her introduction (30), "all the circumstances surrounding the letter, namely, its position in the manuscripts,

the reference to future time in a few of the letters, the speed of transportation of persons, letters, and news, pertinent facts concerning the addressees and persons mentioned in the letters, and events mentioned in the letters as checked by other sources." Where her dating disagrees with the more recent edition, *Gerbert d'Aurillac: Correspondance*, edited and translated (into French) by Pierre Riché and J.-P. Callu, I have followed Lattin; Riché and Callu date this letter to Adalbold as "997 to 999." Neither edition translates Adalbold's query about the volume of a sphere; it can be found in Nicolai Bubnov, ed., *Gerbert d'Aurillac: Opera mathematica (972–1003)*, 302–309 (hereafter cited as "Gerbert, *Opera mathematica*").

3 *"The Scientist Pope":* Ademar of Chabannes, c. 1030, calls Gerbert *philosophus papa* and noted his "scientific talent"; see his *Chronique*, edited by J. Chavanon, 154, and Richard Landes, *Relics, Apocalypse, and the Deceits of History*. Landes and others call Gerbert "The Philosopher Pope" and translate the phrase as "philosophical talent," but since Ademar's time the word "philosophy" has been reduced from "all of learning" to refer to a specific discipline. In the tenth century, the seven liberal arts were a subset of philosophy; now philosophy is a subset of the liberal arts, and science is the larger term. "Great genius" is from Richer of Saint-Remy, *Histoire de France (888–995)*, edited and translated (into French) by Robert Latouche, vol. 2, 51 (hereafter cited as "Richer of Saint-Remy"). Helgaud of Fleury's assessment, c. 1041, is quoted by Riché, *Gerbert d'Aurillac*, 202, as *son incomparable science philosophique*; I have again translated *philosophique* as "scientific" and *science* as "knowledge." "Surpassed his contemporaries," is from Thietmar of Merseburg, *Chronicon* (1013–1018), translated by David Warner in *Ottonian Germany*, 303 (hereafter cited as "Thietmar of Merseburg"). "Acutely intelligent" and "deeply learned" are from Ralph the Bald in Book I of his *Histories* (written before 1030), 27.

5 *first computer:* See "Y1K: Web-Only Supplement to *The Millennium Issue*," *Searcher Magazine*, http://www.infotoday.com/searcher/jan00/y1k.htm. Function "digitally" is the wording of Arno Borst, *The Ordering of Time*, 58. For a sample chronology, see http://www.cyberstreet.com/hcs/museum/chron.htm.

6 *Holy Roman Empire:* The term was not used until the twelfth century, when it was "back-dated" to include Otto the Great as the first Holy Roman Emperor. Otto III would have thought of himself as the Roman Emperor or the Emperor of the West, but for clarity I have retained the modern term. See Henry Mayr-Harting, *Church and Cosmos in Early Ottonian Germany*, 4–5.

Chapter I: A Monk of Aurillac

11 *"occupied in administering":* Quotations from Odo of Cluny's "Life of Saint Gerald of Aurillac" in *St Odo of Cluny*, translated by Gerard Sitwell (hereafter cited as "Odo of Cluny").

13 *Celibacy:* Kathleen G. Cushing describes the problem of "sexually-active clergy" in *Reform and the Papacy in the Eleventh Century*, 98–149.

14 *some pray:* Ascelin of Laon, aka Adalberon de Laon, *Poème au roi Robert*, edited and translated (into French) by Claude Carozzi, 23. I have consistently used his nickname, Ascelin, to distinguish him from his uncle, Adalbero of Reims.

14 *"I do not know":* Gerbert, 236. "Happy day" is Gerbert, 92; translated by Darlington, "Gerbert the Teacher," 457.

15 *Rule of Saint Benedict:* The Rule was modified by Benedict of Aniane in the early 800s from the original of Benedict of Nursia. See Leonard J. Doyle's translation at http://www.osb.org/rb/text/toc.html#toc.

15 *poorest and worst:* Details come from the Rule; Scott G. Bruce, *Silence and Sign Language in Medieval Monasticism*, 177–181; Joan Evans, *Monastic Life at Cluny*, 78–88; and C. H. Lawrence, *Medieval Monasticism*, 112–115. Since Aurillac was "reformed" by Cluny before Gerbert's time, customs at both monasteries were similar; in general, says Evans, life at a Cluniac monastery was "a moderate interpretation of the Benedictine rule, warmed by charity and illumined by beauty in buildings, services, and music; setting psalmody in place of manual labor, cultivating the arts as well as theology; governed by tradition and custom in every detail of the day, yet saved from deadness alike by religious fervor and charitable deeds" (126).

16 *"language of the fingers":* Odo of Cluny, 33. Lawrence describes the monk's sign language, 118–119; Evans supplies "pancakes," *Monastic Life at Cluny*, 88; the signs for "pillow" and the various ideas come from Bruce, 177–181.

17 *alone with his thoughts:* Lawrence, 119–120.

18 *pay compensation:* Lawrence, 69–70.

20 *"he stole them":* Quotations from Bernard of Angers's *Book of Miracles of Saint Foy* in *The Book of Sainte Foy*, translated by Pamela Sheinborn. For saints as fund-raisers, see Patrick Geary, *Furta Sacra*, 24, 71–74.

21 *"by trickery or theft":* The tenth-century *translatio* of Saint Bibanus, as quoted by Geary, 74.

22 *Gerald's original church:* Nicole Charbonnel, "La Ville de Gerbert, Aurillac," in *Gerbert l'Européen*, edited by Nicole Charbonnel and Jean-Eric Iung, 65–70.

23 *list of 63 books:* Lawrence describes the books read at Farfa, 115; Evans those at Cluny in 1042, 101. The 415 books at Reichenau are listed by Rosamund McKitterick, *The Carolingians and the Written Word*, 179–182. For Bobbio, see Jean-François Genest, "Inventaire de la bibliothèque de Bobbio," in Olivier Guyotjeannin and Emmanuel Poulle, eds., *Autour de Gerbert d'Aurillac*, 250–260.

24 *Martianus Capella:* Henry Mayr-Harting, *Church and Cosmos in Early Ottonian Germany*, 21, 197–200, 217–218.

Chapter II: Of the Making of Books There Is No End

25 *"To tell the truth":* Ulrich of Cluny, in Lawrence, *Medieval Monasticism*, 115.

26 *"place it in limewater":* Leandro Gottscher, "Ancient Methods of Parchment-Making," in Marilena Maniaci and Paola F. Munafo, eds., *Ancient and Medieval Book Materials and Techniques*, vol. 1, 47; and the recipe from Bologna, 52.

29 *forty strokes:* Michael Gullick, "How Fast Did Scribes Write?" in Linda L. Brownrigg, ed., *Making the Medieval Book*, 39–58. Another calculation can be found in Philippe Wolff, *The Awakening of Europe*, 59. The number of scribes and artists needed was calculated by William Noel, "The Division of Work in the Harley Psalter," in Brownrigg, 1–16.

30 *bought a lawbook:* McKitterick, *The Carolingians and the Written Word*, 136.

30 *greatest book collectors:* Twelve of Gerbert's extant letters contain requests for books. Citations here are from Gerbert, 53, 54, 168, 50.

32 *"shone with the double light":* Gerbert, 136; translated by Darlington, "Gerbert the Teacher," 458. Eleanor Duckett cites Thangmar's *Life of Bernward* in *Death and Life in the Tenth Century*, 181. Gerbert thanks Raymond twice more; see Gerbert, 140, 230.

34 *"He who cannot write":* Sylvie Fournier, *A Brief History of Parchment and Illumination*, 12; "excessive drudgery," Reviel Netz and William Noel, *The Archimedes Codex*, 220; "Learn to write, boy," John J. Contreni, "The Pursuit of Knowledge in Carolingian Europe," in Richard E. Sullivan, ed., *"The Gentle Voices of Teachers,"* 115. See also Wolff, 55.

35 *"In the name of the fatherland":* From *Life of Saint Boniface;* "male and female mules," from the *Life of Bishop Meinwerk.* Both are cited by G. Tellenbach, *The Church in Western Europe from the Tenth to the Early Twelfth Century*, 25. "Dangling, leaping" is from the notes to *Aelfric's Colloquy*, edited by G. N. Garmonsway, 11n.

35 *"sea without a shore":* Contreni, "The Pursuit of Knowledge in Carolingian Europe," in Sullivan, 121. Contreni is also my source for the outdated textbooks.

35   *"Little Red Riding Hood":* Riché, *Les grandeurs de l'an mil,* 199–200.

36   *the classics:* Riché, *Gerbert d'Aurillac,* 42.

36   *"quench your thirst":* Gerbert, 205; translated by Darlington, "Gerbert the Teacher," 465. "The features of my friend remain fixed in my heart" is from Gerbert, 83, 93; "Equally in leisure and in work" is from Gerbert, 92; "Although really still a learner" is from Gerbert, 45, all in Lattin's translation.

37   *Gerbert's writing:* Riché, *Gerbert d'Aurillac,* 250; Lattin, introduction to *The Letters of Gerbert,* 27–28. Gerbert's "Ciceronian eloquence" is remarked on by Richer of Saint-Remy, vol. 2, 51.

37   *kinds of logic:* Fulbert of Chartres, *The Letters and Poems of Fulbert of Chartres,* translated by Frederick Behrends, xxv–xxvi. Contreni gives the "God is not anywhere" lesson, "The Pursuit of Knowledge in Carolingian Europe," in Sullivan, 123.

38   *"father and light":* Teta E. Moehs, "Gerbert of Aurillac as Link Between Classicism and Medieval Scholarship," in M. Tosi, ed., *Gerberto,* 342. See also Pascale Bourgain, "L'Hommage de Gerbert à Boèce," in Olivier Guyotjeannin and Emmanuel Poulle, *Autour de Gerbert d'Aurillac,* 296–299. "For these cares . . . ," Gerbert, 91.

38   *"so far as life":* Boethius, *The Consolation of Philosophy,* translated by E. Watts, 12.

39   *"practice disputation":* Richer of Saint-Remy, translated by Frederick Behrends in his introduction to *The Letters and Poems of Fulbert of Chartres,* xxx. C. Stephen Jaeger cites Alcuin in *The Envy of Angels,* 32; for "nobility of soul," see 56, 93, 102, 113.

39   *"all things human and divine":* Richer of Saint-Remy, vol. 2, 77. His definition makes it clear that "philosophy" had a much larger meaning for Gerbert than it does today.

40   *"made his escape":* William of Malmesbury, *Gesta Regum Anglorum,* translated by R. A. B. Mynors, R. M. Thomson, and M. Winterbottom, vol. 1, 279 (hereafter cited as "William of Malmesbury").

Chapter III: The Ornament of the World

41   *"The abbot inquired":* Richer of Saint-Remy, translated by Darlington, "Gerbert the Teacher," 460. For Gerbert's letters on Spain, see Gerbert, 115, 117. Borrell's son is quoted by Pierre Riché, *Gerbert d'Aurillac,* 234.

42   *Borrell's Spain:* Paul H. Freedman, *The Diocese of Vic,* 116–118. On the Visigoths' sophisticated code of law, see Richard Fletcher, *The Quest for El Cid,* 13–14.

43   Song of Roland: Maria Rosa Menocal, *The Ornament of the World,* 56–58.

46 *"searched with great zeal":* J. L. Berggren, *Episodes in the Mathematics of Medieval Islam*, 4; he also discusses al-Khwarizmi's work, 6–9. On star tables, see David A. King, "Islamic Astronomy," in Christopher Walker, ed., *Astronomy Before the Telescope*, 143–174.

47 *Al-Khwarizmi's science:* Michael C. Weber argues that it reached Spain during al-Khwarizmi's lifetime; see Thomas Glick, Steven J. Livesey, and Faith Wallis, eds., *Medieval Science, Technology, and Medicine: An Encyclopedia*, 333. Glick says the star tables, at least, had arrived "in the mid-ninth century," in *From Muslim Fortress to Christian Castle*, 47. According to Georges Hourani, "Indian numerals and the Arabic zero as explained by Khwarizmi were known in Andalusia before 859"; see "The Early Growth of the Secular Sciences in Andalusia," 148.

48 *magnificent land:* Muhammad Ibn Hauqal, *Configuration de la terre (Kitab surat al-Ard)*, edited and translated (into French) by J. H. Kramers and G. Wiet. For Hasdai ibn Shaprut's letter, see Hasdai ibn Shaprut, "The Epistle of R. Chisdai, son of Isaac . . . " in *Jewish Travellers*, translated by Elkan Nathan Adler, 22–32. For the calendar, see the anonymous work *Le Calendrier de Cordoue*, edited by R. Dozy and translated (into French) by Ch. Pellat.

49 *paper-making:* The name "the papermaker" is cited by Glick, *Islamic and Christian Spain*, 279. Menocal locates a tenth-century paper mill near Valencia, 34, and describes "astonishing" Cordoba, 32. Hourani discusses the 400,000 (or 40,000) books in the caliph's library, 148.

50 *"ornament of the world":* Menocal translates from Hrosvit of Gandersheim's poem "Pelagius," 12. For the complete poem, see Hrotsvit of Gandersheim, *Hrotsvit of Gandersheim: A Florilegium of Her Works*, translated by Katharina Wilson, 29–40. Stephen C. McCluskey notes Racemundo's contributions to the Calendar of Cordoba in *Astronomies and Cultures in Early Medieval Europe*, 166–170. Menocal discusses the *dhimma*, 72ff.

51 *"group of doctors":* Hourani translates Ibn Juljul's anecdote, 155–156.

52 *"the numbers and the figures":* Menso Folkerts, "The Names and Forms of the Numerals on the Abacus in the Gerbert Tradition." David Juste brings up the number of manuscripts of the astrolabe treatise to argue that Gerbert did not write it; see *Les Alchandreana primitifs*, 219–294.

52 *"surpassed Ptolemy":* William of Malmesbury, vol. 1, 279. Richer of Saint-Remy is cited by Darlington in "Gerbert the Teacher," 460. Ademar de Chabannes is quoted by Marco Zuccato, who develops the idea of "Gotmar's circle" in "Gerbert of Aurillac and a Tenth-Century Jewish Channel for the Transmission of Arabic Science to the West," 750–752.

56 *Abbot Garin:* André Bonnery, *L'abbaye Saint-Michel de Cuixà*; he quotes Miro, 9. Also Pierre Riché, *Les grandeurs de l'an mil*, 116–117. For his let-

ters, see Gerbert, 63, 70, 73, 91. Zuccato identifies Hasdai as "Joseph the Wise" in "A Tenth-Century Jewish Channel," 754.

57 *school at Ripoll:* Joseph Pijoan, "Oliba de Ripoll (971–1046)," and Freedman, 25–26. On the manuscript Ripoll 225, see Paul Kunitzsch, "Al-Khwarizmi as a Source for the *Sententie astrolabii*," in D. A. King and G. Saliba, eds., *From Deferent to Equant*, 227. The libraries in Ripoll and Vic are compared by Michel Zimmerman, "La Catalogne de Gerbert," in Nicole Charbonnel and Jean-Eric Iung, eds., *Gerbert l'Européen*, 84.

58 *Lobet:* Gerbert, 69. Lattin identifies Lobet in "Lupitus Barchinonensis."

58 *Vigila:* See Juste, *Les Alchandreana primitifs*, 246; Marcel Destombes, "Un astrolabe carolingien et l'origine de nos chiffres arabes," 9; and Georges Ifrah, *The Universal History of Numbers*, 362.

58 *Miro Bonfill:* See Juste, *Les Alchandreana primitifs*, 242–247. That "the wisest scholar" could have been Maslama of Madrid is my speculation, not Juste's. For the letter, see Gerbert, 70. On the Elne stone, see Flavio Nuvolone's works, particularly "Elna e l'iscrizione attribuita a Gerberto d'Aurillac: Gerberto si recorda del vescovo e delle martiri?"

Chapter IV: The Schoolmaster of Reims

63 *mission to Rome:* John of Salerno, "The Life of Saint Odo of Cluny," translated by Gerard Sitwell, 60–61; Richer of Saint-Remy, vol. 2, 229. For Gerbert's meeting with the pope, see Marco Zuccato's translation in "Gerbert of Aurillac and a Tenth-Century Jewish Channel for the Transmission of Arabic Science to the West," 748.

65 *"Socratic disputations":* Gerbert, 82.

65 *courtiers:* Stephen C. Jaeger, *The Origins of Courtliness*, 31, 45.

65 *"directed toward him":* Richer of Saint-Remy, translated by Darlington, "Gerbert the Teacher," 463. "Divinity wished to illuminate Gaul" is the translation of Jason Glenn in *Politics and History in the Tenth Century*, 63.

66 *Adalbero:* See Richer of Saint-Remy, vol. 2, 29–49. For the letters, see Gerbert, 188, 200, 221. Glenn, 25–48, describes Adalbero's reforms and his rebuilding of the Reims cathedral; he also translates some of Richer of Saint-Remy's descriptions of the monk's clothing, 74. Pierre Riché gives Adalbero's biography in *Gerbert d'Aurillac*, 36; he discusses "prince-bishops" in *Les grandeurs de l'an mil*, 137, 141–142. Geoffrey Koziol calls Adalbero an "imperial mole" and explains the crowns over the altars in *Begging Pardon and Favor*, 113–116. In *The Diocese of Vic*, Freedman describes a cathedral as a "club" and explains the difference between monks and canons, 44.

68 *proto-university:* Lists of Gerbert's students are found in Glenn, 64; Anna Marie Flusche, *The Life and Legend of Gerbert of Aurillac*, 30; and Darlington, "Gerbert the Teacher," 473. Darlington cites *The Life of King Robert*, written

c. 1040. The wandering scholars are quoted by Loren C. MacKinney, *Bishop Fulbert and Education at the School of Chartres*, 12. For his letters on teaching, see Gerbert, 85, 59, 296, 44, 189, 90.

69 *Constantine:* "Sweet solace," Gerbert, 45. Jaeger quotes the ode in *The Envy of Angels*, 56–59. Constantine is credited with preserving Gerbert's letter collection by Riché and Callu in their introduction to Gerbert's *Correspondences*, xi–xii. Charles Burnett discusses pen names in "King Ptolemy and Alchandreus the Philosopher," 333. For Constantine's biography, see F. M. Warren, "Constantine of Fleury, 985–1014"; and Thomas Head, *Hagiography and the Cult of Saints*, 217–240, "Letaldus of Micy and the Hagiographic Traditions of the Abbey of Nouaillé," and "The Development of the Peace of God in Aquitaine (970–1005)."

70 *friendship:* See Riché, *Gerbert d'Aurillac*, 251–254; Giles Constable, "Dictators and Diplomats in the Eleventh and Twelfth Centuries"; Lattin, "The Letters of Gerbert," in M. Tosi, ed., *Gerberto*, 311–331; Riché, "Le siège de Laon de 988," in Olivier Guyotjeannin and Emmanuel Poulle, *Autour de Gerbert d'Aurillac*, 126–133; Focillon, *The Year 1000*, 129, 134; and J.-P. Callu, "Les Mots de Gerbert," in Nicole Charbonnel and Jean-Eric Iung, eds., *Gerbert l'Européen*, 151. For his letters on friendship, see Gerbert, 92, 90 (I have amended Gerbert's reference to "Tully" to the more familiar "Cicero"), 140, 163, 180, 184, 207, 196, 199, 151. Jaeger describes true friendship in *Envy of Angels*, 104.

72 *"quantities of sweat":* Richer of Saint-Remy, vol. 2, 51–81; these translations are Darlington's in "Gerbert the Teacher," 464–472. Richer's biography, and the close reading of his autograph manuscript, are Glenn's, especially 127–129, 176. See also Riché, *Gerbert d'Aurillac*, 40; and Juste, *Les Alchandreana primitifs*, 255.

75 *What did Gerbert know:* Arianna Borrelli discusses the oral transmission of Arabic science in *Aspects of the Astrolabe*; also Marco Zuccato (personal communication).

Chapter V: The Abacus

79 *"notorious Gerbert":* Walter Map, *De nugis curialium*, edited and translated by M. R. James, C. N. L. Brooke, and R. A. B. Mynors, 351.

79 *Gerbert's abacus:* Burnett, "The Abacus at Echternach in ca. 1000 AD" and "Algorismi vel helcep decentior est diligentia: The Arithmetic of Adelard of Bath and His Circle"; he discusses Gerbert's students at Mettlach in "The Abacus at Echternach," 104.

80 *"you alone bear burdens":* For his letters, see Gerbert, 117, 111, 102, 114.

81 *first calculator:* Tim Roufs, "Inter-facing the Inevitable: Appropriate Technology in the 21st Century." Osmo Pekonen accuses Gerbert of "showing

off with his supercomputer," in "Gerbert d'Aurillac: Mathematician and Pope," 69.

82 *"last number of the abacus":* Gerbert, 290.

82 Book of the Abacus: For his letter to Constantine, see Gerbert, 45; for the rules, see Gerbert, *Opera mathematica,* 6–22. The discussion of "intervals" is from Gillian R. Evans, "*Difficillima et Ardua:* Theory and Practice in Treatises on the Abacus, 950–1150," 34. On Ralph of Laon, see Evans, "Schools and Scholars: The Study of the Abacus in English Schools c. 980–c. 1150," 81.

86 *"gesticulations of dancers":* Alexander Murray, *Reason and Society in the Middle Ages,* 164.

86 *"When you say one":* Georges Ifrah cites Bede's instructions, including how to give warnings, in *From One to Zero,* 72–77, and Arabic insults, 78.

87 *Roman numerals:* David King derives Roman numerals from finger counting through Greek letters in *The Ciphers of the Monks,* 281–290.

88 *Arabic numerals:* Menso Folkerts, "The Names and Forms of the Numerals on the Abacus in the Gerbert Tradition." See also King, *The Ciphers of the Monks,* 309–316. Otto B. Bekken cites Severus Sebokt and al-Khwarizmi in the "Algorismus of *Hauksbók:* An Old Norse Text of 1310 on Hindu-Arabic Numeration and Calculation," http://home.hia.no/~ottob/. Ifrah describes the practical handbook of al-Buzjani in *The Universal History of Numbers,* 548.

89 *"dry" and "dull":* Evans discusses Gerbert's influence in "*Difficillima et Ardua*" and in "Schools and Scholars," where she quotes William of Malmesbury, 72. Burnett defines "abacist" and *abaci doctor* in "Abbon de Fleury *Abaci Doctor,*" in Barbara Obrist, ed., *Abbon de Fleury,* 129–139. Lattin notes that abacists were called "gerbercists" (she spells it "girbercists") in the introduction to *The Letters of Gerbert,* 19.

90 *Adelard:* Burnett, "Algorismi vel helcep." For the connection of Laon to Gerbert, see Evans, "Schools and Scholars," 74.

90 *took much longer:* See J. L. Berggren, *Episodes in the Mathematics of Medieval Islam,* 29–39; Burnett, "Arabic Numerals," in Thomas Glick, Steven J. Livesey, and Faith Wallis, eds., *Medieval Science, Technology, and Medicine,* 39–40, Burnett, "The Translating Activity in Medieval Spain," in *Magic and Divination in the Middle Ages,* 1040; and King, *The Ciphers of the Monks,* 313–315.

Chapter VI: Math and the Mind of God

93 *strangely transformed:* John J. Contreni, "The Pursuit of Knowledge in Carolingian Europe," in Richard E. Sullivan, ed., "*The Gentle Voices of Teachers,*" 117.

94   *massive times table:* Abbo of Fleury, *Commentary on the Calculus of Victo-rius of Aquitaine,* edited by A. M. Peden, xv–xxxvi.

94   *"your sagacity":* Gerbert, 44; "half-educated philosopher," 46. Lattin iden-tifies him as Abbo, 46n. See also Marco Mostert, "Gerbert d'Aurillac, Abbon de Fleury, et la Culture de l'An Mil," in Nuvolone, ed., *Gerberto d'Aurillac: Da abate di Bobbio a Papa dell'anno 1000,* 405.

94   *Abbo:* Vezin points out that Abbo's and Gerbert's "curiosities" coincided in "La Production et la circulation des livres dans l'Europe du Xe siècle," in Nicole Charbonnel and Jean-Eric Iung, eds., *Gerbert l'Européen,* 215. John Nightingale compares Ramsey and Fleury in "Oswald, Fleury, and Con-tinental Reform," in Nicholas Brooks and Catherine Cubitt, eds., *St Os-wald of Worcester,* 27, 44. Thomas Head cites Abbo's medieval biographer, Aimo of Fleury, in *Hagiography and the Cult of Saints,* 239–240. Pierre Riché describes the reform movement, *Les grandeurs de l'an mil,* 114; as does Richard Landes, *Relics, Apocalypse, and the Deceits of History,* 29. Robert F. Berkhofer discusses charters, *Day of Reckoning,* 73. Teta E. Moehs describes Fleury's privileges, *Gregorius V, 996–999,* 21, 47.

96   *Abbo studied at Reims:* Aimo of Fleury's comment is translated by Jason Glenn, *Politics and History in the Tenth Century,* 62. Charles Burnett iden-tifies Gerbert as Abbo's teacher in "King Ptolemy and Alchandreus the Philosopher," 332; as does David Juste, "La sphère planétaire du ms. Vat-ican, BAV, Pal. lat. 1356 (XIIe siècle)," 217n. C. H. Lawrence quotes Abbo on writing, *Medieval Monasticism,* 117. On Abbo's scholarship, see Peden's introduction to Abbo's *Commentary,* xiii–xiv; Ron B. Thomson, "Two Astronomical Tractates of Abbo of Fleury," 113–116; Barbara Obrist, "Les tables et figures abboniennes dans l'histoire de l'iconographie des recueils de comput," in Obrist, ed., *Abbon de Fleury,* 141–186; and Gillian R. Evans, "Schools and Scholars," 73–75. Wesley M. Stevens calls Byrthferth's work "disappointing" in *Cycles of Time and Scientific Learning in Medieval Europe,* 138–139. Discussing Abbo's use of Boethius's *On Arithmetic,* Peden notes, "It is the mark, perhaps, of a text well-absorbed, even if only the more elementary sections," xxxiv.

98   *"Take number away":* Isidore of Seville, quoted by Faith Wallis, "'Number Mystique' in Early Medieval Computus Texts," in Teun Koetsier and Luc Bergmans, eds., *Mathematics and the Divine,* 187.

98   *most common math problem:* Stevens, *Cycles of Time,* 44–45.

100 *800 Anno Domini:* See Landes, *Relics, Apocalypse, and the Deceits of History,* 290. Hraban Maur's *computus* is discussed by Wesley M. Stevens, "Com-potistica et Astronomica in the Fulda School," 32–34. For Abbo's recal-culations, see Arno Borst, *The Ordering of Time,* 52–54; Johannes Fried, "Awaiting the End of Time Around the Turn of the Year 1000," in Richard Landes, Andrew Gox, and David C. Van Meter, eds., *The Apocalyptic Year*

1000, 21, 37; and Landes, "The Fear of an Apocalyptic Year 1000," in the same work, 250–252.

100 *"unchangeable and true"*: Abbo of Fleury, summarized by Peden, xvii–xviii.

101 *"I did not restrict myself"*: Boethius, *De institutione arithmetica*, in *Boethian Number Theory*, edited and translated by Michael Masi, 67. "The original pattern" is from Joseph V. Navari, "The Leitmotiv in the Mathematical Thought of Gerbert of Aurillac," 140–141.

101 *Boethius's* On Arithmetic: Gerbert's gold and purple copy is described by Rosamond McKitterick, "Charles the Bald (823–877) and His Library: The Patronage of Learning," 31. For his letters, see Gerbert, 296, 39. Menso Folkerts finds the problems "completely irrelevant for practical calculation" in *The Development of Mathematics in Medieval Europe*, Chap. 1, 16. For the meaning of the exercises, see Alison White, "Boethius in the Medieval Quadrivium," in Margaret Gibson, ed., *Boethius: His Life, Thought, and Influence*, 169–170; Navari, 141–142; and Wallis, in Koetsier and Bergmans, 196.

103 *organ pipes:* See Flusche, *The Life and Legend of Gerbert of Aurillac*, 147–156; Christian Meyer, "Gerbertus musicus: Gerbert et les fondements du système acoustique," in Charbonnel and Iung, 183–192.

104 musica humana: C. Stephen Jaeger describes moral music in *Envy of Angels*, 165.

105 *"sweet mixture"*: Bukofzer, in "Speculative Thinking in Mediaeval Music," discusses the history of chant, 168–172; he cites the "deeper and divine reason" controlling harmony, from the anonymous *Musica enchiriadis* (c. 860), 173.

105 *master of* musica: Richer of Saint-Remy, vol. 2, 53–59. "But the difficulties" is my translation. "Music, previously unfamiliar to the Gauls," is that of Glenn, 50; he also discusses the evidence that Richer was a cantor, 49–52.

106 *His solution:* Constantino Sigismondi, "Gerberto e la misura delle canne d'organo," 392–393.

106 *"full of accurate observations"*: Quotes from Gerbert's geometry are translated by Navari, 141, 147. See also Menso Folkerts, "The Importance of the Pseudo-Boethian Geometria During the Middle Ages," in *The Development of Mathematics in Medieval Europe*, Chap. 7, 190–201. Stevens cites Folkerts's work as proof of the "quite unexpected amount" of Euclid, in "Compotistica et Astronomica," 40–42. On Gerbert's letter to Adalbold (Gerbert, 299), see G. A. Miller, "The Formula ½a (a + 1) for the Area of an Equilateral Triangle" and "Gerbert's Letter to Adelbold."

111 *experimental science:* See Folkerts, "The Importance of the Latin Middle Ages for the Development of Mathematics," in *The Development of Mathematics in Medieval Europe*, Chap. 1, 4; see also his "Review of *De*

Verhandeling over de Cirkelkwadratuur van Franco van Luik van Omstreeks 1050 by A. J. E. M. Smeur," 272–274. Loren C. MacKinney continues the story of Rodolf and Ragimbold, *Bishop Fulbert and Education at the School of Chartres*, 14–15, 29–30; as does Michael Mahoney, "Mathematics," in David Lindberg, ed., *Science in the Middle Ages*, 149.

111　*prayer book:* Reviel Netz and William Noel, *The Archimedes Codex.*

Chapter VII: The Celestial Sphere

113　*"giant silvery cloud":* Joe Sharkey, "Helping the Stars Take Back the Night," *New York Times*, Aug. 30, 2008. For how Gerbert's peers considered the stars, see C. Stephen Jaeger, *Envy of Angels*, 175; David King, *Ciphers of the Monks*, 355; Stephen C. McCluskey, *Astronomies and Cultures in Early Medieval Europe*, 3; and Ron B. Thomson, "Two Astronomical Tractates of Abbo of Fleury," 113–133.

114　*Gregory of Tours:* Arno Borst, *The Ordering of Time*, discusses temporal and equinoctial hours and the use of the alarm *glocke*, 31, 42.

115　*"it would take too long":* Richer of Saint-Remy, translated by Darlington, "Gerbert the Teacher," 467.

115　*"We have sent no sphere":* Gerbert, 172. "Your good will," Gerbert, 184.

116　*Reims and Trier:* See Thomas Head, "Art and Artifice in Ottonian Trier," 65–82; Dominique Alibert, "Majesté ottonienne: L'hommage des nations à l'empereur," in Olivier Guyotjeannin and Emmanuel Poulle, eds., *Autour de Gerbert d'Aurillac*, 82–87; and Pierre Riché, *Les grandeurs de l'an mil*, 191.

116　*"Continual torrents":* Gerbert, 149. He mentions the cross in two other letters; see Gerbert, 145, 147.

117　*Gerbert's student:* For Remi, see Darlington, "Gerbert the Teacher," 470–471. For Leofsin, see Charles Burnett, "The Abacus at Echternach in ca. 1000 AD," 104.

118　*"the Celestial Sphere":* Bruce Eastwood cites Martianus Capella in *Ordering the Heavens*, 189. For Al-Battani's "egg" and its appearance in Catalonia, see Julio Samso, "Battani, al-," in Thomas Glick, Steven J. Livesey, and Faith Wallis, eds., *Medieval Science, Technology, and Medicine*, 79–80. Al-Sufi's silver sphere is described by E. S. Kennedy and Marcel Destombes in "Introduction to *Kitab al'Amal bil Asturlab*," in E. S. Kennedy et al., eds., *Studies in the Islamic Exact Sciences*, 405–406.

118　*Gerbert may have learned:* On Gerbert's technique, see David Juste, "La sphère planétaire du ms. Vatican, BAV, Pal. lat. 1356 (XIIe siècle)," 205–221; and Marco Zuccato, "Gerbert's Islamicate Celestial Globe," 167–188. Emilie Savage-Smith quotes the Arabic text translated in Spain, *Islamicate Celestial Globes*, 81–82. For the 1518 printed text, see R. Lorch, "The *Sphaera Solida* and Related Instruments," 156.

119 *two ways of thinking:* See Savage-Smith, 3, and Wesley M. Stevens, "The Figure of the Earth in Isidore's *De natura rerum*," 275–277. Martianus Capella called the planets "confusers"; see Eastwood, *Ordering the Heavens*, 323. "There is no inconstancy in divine acts" is Calcidius, quoted by Eastwood, "Calcidius's Commentary on Plato's *Timaeus* in Latin Astronomy of the Ninth to Eleventh Centuries," 176.

120 *"something divine":* Richer of Saint-Remy, translated by Zuccato, "Gerbert's Islamicate Celestial Globe," 169. "Differ from organ pipes" is from Gerbert, 36. "So well contrived" is from Richer of Saint-Remy, translated by Darlington, "Gerbert the Teacher," 467.

121 *most sophisticated astronomical instrument:* See Juste, "La sphère planétaire," 205–221. According to Richer's numbering system, which Juste uses, the armillary sphere is the third sphere, not the fourth.

123 *eclipses:* See Thietmar of Merseburg, 161; Liudprand of Cremona, translated by Francis Wright, *The Works of Liudprand of Cremona*, 177, 275; and Ralph the Bald, 211, 213, 241, 245. The *Life of Heraclius* is quoted by Bruce Eastwood, *The Revival of Planetary Astronomy in Carolingian and Post-Carolingian Europe*, 250.

124 *globus:* Stevens, "The Figure of the Earth in Isidore's *De natura rerum*," 275–277. Saint Augustine's *The Literal Meaning of Genesis*, Book 1, chapter 19, is translated by J. H. Taylor. On Lactantius and Cosmas, see Jeffrey Burton Russell, *Inventing the Flat Earth*, 32–35; also Alain Touwaide, "Kosmas Indikopleustes," in Glick et al., 302–303.

126 *Lady Geometry:* On the glosses to Martianus Capella, see Natalia Lozovsky, *"The Earth is Our Book,"* 114–130.

126 *"golden apple":* Ralph the Bald, 39.

126 *Maps of the world:* See Patrick Gautier Dalché, "Mappemonde dessinée à Fleury vers l'an mil," in Guyotjeannin and Poulle, 2–5; Lozovsky, 114–130; and Stevens, "The Figure of the Earth in Isidore's *De natura rerum*," 268–277.

128 *Antipodes:* Saint Augustine's *The City of God*, Book 16, chapter 9, is translated by Marcus Dods, Christian Classics Ethereal Library at Calvin College, http://www.ccel.org/ccel/schaff/npnf102.iv.XVI.9.html.

128 *Flat Earth Error:* Russell, especially 65–73; for Columbus's alternative history, see 3–11. Quotations from Washington Irving are from *Columbus: His Life and Voyages*, vol. 1, 33–38. Russell found the Flat Earth Error in a 1983 textbook for fifth-graders; a 1982 textbook for eighth-graders; the 1960, 1971, and 1976 editions of a college textbook, *A History of Civilization*; and in the bestselling 1983 book *The Discoverers*, by the former Librarian of Congress, Daniel Boorstin. I have since found it in Haraldur Sigurdsson's 1999 book, *Melting the Earth: The History of Ideas on Volcanic Eruptions*, 71.

Chapter VIII: The Astrolabe

133   *Ptolemy:* For the Arabic folktale, see David King, "Astronomical Instruments Between East and West," 146. For the Latin, see Charles Burnett, "King Ptolemy and Alchandreus the Philosopher," 340–341.

133   *1,760 uses:* Abd al-Rahman al-Sufi (from Shiraz, c. 965) is cited by David King, *Astrolabes and Angels*, 199.

134   *magical instrument:* Arianna Borrelli discusses Gerbert's role in the oral transmission of knowledge about the astrolabe in *Aspects of the Astrolabe*; also Marco Zuccato (personal communication).

134   *surpassed Ptolemy:* William of Malmesbury, 279–289. On Michael Scot, see Lynn Thorndike, *Michael Scot*, 93–94. "It should be noted" is translated from the French of Poulle, "Naissance de la légende scientifique: Note sur l'autorité des traités de l'astrolabe," in Olivier Guyotjeannin and Emmanuel Poulle, eds., *Autour de Gerbert d'Aurillac*, 343.

135   *"particularly skilled":* Thietmar of Merseburg, 303. For translations of *horologium*, see Roland Allen, "Gerbert, Pope Silvester II," 633; Gerd Althoff, *Otto III*, 69; Brigitte Bedos-Rezak, "Review of *Autour de Gerbert d'Aurillac*," 529; Bruce Eastwood, *The Revival of Planetary Astronomy*, 253; Anna Marie Flusche, *The Life and Legend of Gerbert of Aurillac*, 59; David Juste, "La sphère planétaire du ms. Vatican, BAV, Pal. lat. 1356 (XIIe siècle)," 208; Stephen C. McCluskey, *Astronomies and Cultures in Early Medieval Europe*, 176; and Poulle, "Gerbert Horloger!" in Guyotjeannin and Poulle, 365–367.

136   *"actual numbers":* Gerbert, 45.

136   *Synesius:* Translation by A. Fitzgerald on *Livius.org: A Website on Ancient History*, edited by Jona Lendering, http://www.livius.org/su-sz/synesius/synesius_astrolabe_3.html.

137   *"rotatable star map":* The description is from the Deutsches Museum; "flat model" is from Marcel Destombes, "Un astrolabe carolingien et l'origine de nos chiffres arabes," 10; "analogue computer" is from E. S. Kennedy and Marcel Destombes, in E. S. Kennedy et al., eds., *Studies in the Islamic Exact Sciences*, 405, and also J. L. Berggren, *Episodes in the Mathematics of Medieval Islam*, 173; "two-dimensional model" is from King, "Astronomical Instruments Between East and West," 145–146.

139   *Rodolf:* The letters of Rodolf and Ragimbold are edited by Paul Tannery in *Mémoires scientifiques*, vol. 5, 229–303. McCluskey discusses them in *Astronomies and Culture*, 177; see also David King, *The Ciphers of the Monks*, 370. Translation by Monica Otter, personal communication.

140   *Pope Sylvester's:* See David King, *In Synchrony with the Heavens*, 489; Robert T. Gunther, *Astrolabes of the World*, 230–232.

140 *Destombes astrolabe:* See Destombes, "Un astrolabe carolingien"; King, *In Synchrony with the Heavens*, 205, 209. For Gerbert's letters on Catalan independence, see Gerbert, 115, 153.

143 *Fulbert:* Fulbert of Chartres, *The Letters and Poems of Fulbert of Chartres*, translated by Frederick Behrends, xxv–xxvii, 260–261; also Burnett, "King Ptolemy and Alchandreus the Philosopher," 334–335.

144 *book on* astrology: David Juste, *Les Alchandreana primitifs*, 219–261.

146 *"Nectanabo":* Adalbero (Ascelin) of Laon, *Poème au roi Robert*, edited and translated (into French) by Claude Carozzi, 8–13.

146 *"house of Mercury":* Juste's translation of Gerbert's epitaph for Duke Ferry of Upper Lorraine, the husband of Beatrice (who would later play a role in Gerbert's career); Juste, *Les alchandreana primitifs*, 256. See also Gerbert, 120. Lattin dates the death of Ferry to May 18, following church necrologies; she notes that the sun entered Gemini (the "house of Mercury") on May 16 (see Gerbert, 120n). The astrological lore permits both readings.

146 *"shattered by the* phisicis": Gerbert, 149.

147 *"I passed by Chartres":* Loren C. Mackinney, *Bishop Fulbert and Education at the School of Chartres*, 14.

148 *"remarkable musician":* From his obituary in the *Annals of Saint-Maixent of Micy*, cited by Thomas Head, "Letaldus of Micy and the Hagiographic Traditions of the Abbey of Nouaillé," 264. Also, Gerbert, 140.

148 *"Ascelin the German":* See Burnett, "King Ptolemy and Alchandreus the Philosopher," 343–354.

148 *beloved Constantine:* Head reconstructs Constantine's biography and distinguishes him from the second Abbot Constantine of Micy in "Letaldus of Micy and the Hagiographic Traditions of the Abbey of Nouaillé," 253–267.

Chapter IX: The Abbot of Bobbio

153 *"interrupted for a time":* Gerbert, 61.

153 *"seemed capable":* Richer of Saint-Remy, vol. 2, 65–67.

155 *"purple-born":* Thietmar of Merseburg, 102. Romily Jenkins says Thietmar misunderstood: Theophanu was "only" the niece of the reigning emperor, John Tzimisces, but her parents were the former Emperor Romanus II and Empress Theophanu; see *Byzantium: The Imperial Centuries*, 293–295. Jacqueline Lafontaine-Dosogne notes that all images of Otto II and Theophanu represent them as equals, in "The Art of Byzantium and Its Relation to Germany in the Time of the Empress Theophano," in Adelbert Davids, ed., *The Empress Theophano*, 212.

155 ingenio facundam: Karl Leyser, "*Theophanu divina gratia imperatrix augusta:* Western and Eastern Emperorship in the Later Tenth Century," in

Davids, 1–27. K. Ciggaar records the opinion of Albert of Metz ("un-pleasantly talkative") and the vision of Theophanu damned, in "Theo-phano: An Empress Reconsidered," in Davids, 49–63. Pierre Riché finds "moderation" and "good manners" in Thietmar's description, in *Les grandeurs de l'an mil*, 95.

155 *"Greek woman":* Odilo of Cluny's "The Epitaph of Adelheid," translated by Sean Gilsdorf, *Queenship and Sanctity*, 133–134; she is also "that Greek empress." For knowledge of Greek in the tenth century, see Henry Mayr-Harting, *Church and Cosmos in Early Ottonian Germany*, 52–57, 144, 198; Florentine Mutherich, "The Library of Otto III," in Peter Ganz, ed., *The Role of the Book in Medieval Culture*, 15–17; and Hrotsvit of Gandersheim, *Hrosvit of Gandersheim: A Florilegium of Her Works*, translated by Katha-rina Wilson, 6–7.

156 *"two prongs":* Saint Peter Damian's diatribe on the fork is quoted by Davids, "Marriage Negotiations Between Byzantium and the West and the Name of Theophano in Byzantium," in Davids, 110. See also John Julius Norwich, *Byzantium*, 257–259.

156 *not the man:* Thietmar of Merseburg, 126. For Otto's visit to Saint Gall, from the *Life of Meinwerk*, see Francis Tschan, *Saint Bernward of Hildesheim*, vol. 2, 19n.

156 *Otric and Gerbert:* Richer of Saint-Remy, vol. 2, 65–81; Thietmar of Merseburg, 136–139. See also Pierre Riché, "L'enseignement de Gerbert," in M. Tosi, *Gerberto*, 62. Riché identifies the eyewitness as Abbot Adso of Montier-en-der, with whom Gerbert shared books, in *Les grandeurs*, 125. Such scholarly debates showed "how huge a geographical triangle was joined up" by a king's patronage of learning, according to Henry Mayr-Harting, *Church and Cosmos in Early Ottonian Germany*, 55, 144.

158 *Miro:* Riché finds Gerbert, Adalbero, and Miro at the Easter synod of 981. See *Gerbert d'Aurillac*, 63.

158 *Bobbio:* For his letters about Bobbio, see Gerbert, 61, 54, 57, 56, 49, 51, 13, 52, 56 (emphasis added). Jean-François Genest describes Bobbio's li-brary in "Inventaire de la bibliothèque de Bobbio," in Olivier Guyotjean-nin and Emmanuel Poulle, eds., *Autour de Gerbert d'Aurillac*, 250–261.

160 *"little books":* Pierre Racine, "Le Monastère de Bobbio et le Monde Féo-dal au Temps de Gerbert," in Flavio G. Nuvolone, ed., *Gerberto d'Auril-lac: Da abate di Bobbio a Papa dell'anno 1000*, 270–276; Riché, *Gerbert d'Aurillac*, 63–71, and *Les grandeurs*, 32–37, 52. "The illustrious Count Oberto" is described by Liudprand of Cremona, as quoted by Racine, 280.

161 *hold on Italy:* Widukind is quoted by Eleanor Duckett, *Death and Life in the Tenth Century*, 70–71. See Thietmar of Merseberg, 93–94, 135, 143–146; for Odilo of Cluny's version, see Gilsdorf, 6–7, 130–131. See also

Paolo Cammarosano, "Gerbert et l'Italie de son temps," in Nicole Charbonnel and Jean-Eric Iung, eds., *Gerbert l'Européen*, 109.

165 *Carmen Figuratum:* Flavio G. Nuvolone, "La Presenza delle Cifre Indo-Arabe nel *Carmen Figurato* di Gerberto: Una discussione" and "Gerbert d'Aurillac et la politique impériale ottonienne en 983: une affaire de chiffres censurée par les moines?" Also Clyde W. Brockett, "The Frontispiece of Paris, Bibl. Nat. Ms. Lat. 776: Gerbert's Acrostic Pattern Poems." For his letter, see Gerbert, 54.

169 *fled to Pavia:* For his letters, see Gerbert, 66, 59, 67, 60, 65, 168, 61, 70, and 68.

171 *Otto III:* Thietmar of Merseberg, 147–156; see also Gerd Althoff, *Otto III*, 31–53. On Lorraine, see Richer of Saint-Remy, vol. 2, 83–91; also Duckett, 101. For Gerbert's letters in support of Theophanu, see Gerbert, 62, 67, 71, 72, 86, 91, 92, 61, 140, 115.

Chapter X: Treason and Excommunication

177 *"The world shudders":* Gerbert, 87. The following letters are from Gerbert, 76, 79, and 95.

178 *Hugh Capet:* Richer of Saint-Remy, vol. 2, 107–115. Karl Leyser critiques Richer's account in *Communications and Power in Medieval Europe*, 170. See also Jim Bradbury, *The Capetians*, 41–82; Elizabeth M. Hallam, *Capetian France*, 20–24, 67–69.

182 *"Louis Do-Nothing":* Richer of Saint-Remy, vol. 2, 117–121.

182 *Adalbero's position:* Gerbert, 99, 97, 98, 106, 107, 108.

184 *magnificent funeral:* Gerbert, 117. He also wrote Lothar's epitaph, Gerbert, 120; according to Pierre Riché, it lacks any sense of sadness, *Gerbert d'Aurillac*, 94. Richer of Saint-Remy, vol. 2, 143, describes the funeral but does not mention Gerbert's involvement. See also Geoffrey Koziol, *Begging Pardon and Favor*, 119–121.

184 *Charles:* Richer of Saint-Remy reports on the conflict for the throne in vol. 2, 145–181. For his letters, see Gerbert, 135, 138, 160, 161, 163, 165. Leyser, 175–176, points out the economic reasons against Charles, *Communications and Power*, 175–176. Riché discusses Charles's continuing support in *Les grandeurs de l'an mil*, 75.

187 *Adalbero of Reims fell ill:* Gerbert, 188, 200, 189.

188 *Arnoul:* Richer of Saint-Remy, vol. 2, 183–225, 231–267. Gerbert's version is presented in *The Acts of Saint-Basle*; see Riché, *Gerbert d'Aurillac*, 126–140; C. Carozzi, "Gerbert et le concile de St-Basle," in M. Tosi, *Gerberto*, 661–676. For his letters, see Gerbert, 186, 192, 196, 202, 206, 209, 218, 216, 230, 236. Jason Glenn, *Politics and History in the Tenth Century*, analyzes the differences between the two versions, 98–127. Koziol dis-

cusses Arnoul's act of prostration, 1–5. On Abbo's "clamor," see Barbara
H. Rosenwein, Thomas Head, and Sharon Farmer, "Monks and Their Enemies," 771, 780; Patrick Geary, *Furta Sacra*, 23. Arnoul and Arnulf are
the same name; to keep the archbishop of Reims straight from the bishop
of Orleans, I have arbitrarily chosen the French spelling for one and the
German spelling for the other.

193 *power of the pope:* Riché quotes Arnulf's diatribe and the papal legate's
reply in *Gerbert d'Aurillac*, 130–134. The bribe of the white horse is described by G. Tellenbach, *The Church in Western Europe from the Tenth to
the Early Twelfth Century*, 88.

194 *excommunicated:* Gerbert, 282, 265.

Chapter XI: The Legend of the Last Emperor

200 *Christian empire:* Henry Mayr-Harting discusses the imperial dreams of
Ottos I and II in *Church and Cosmos in Early Ottonian Germany*, 143. On
the Legend of the Last Emperor, see Matthew Gabriele, "Otto III, Charlemagne, and Pentecost A.D. 1000," in Frassetto, *The Year 1000*, 111–123.

200 *"fiery glowing eyes":* Widukind's description of Otto I is quoted by Karl
Leyser, *Rule and Conflict in an Early Medieval Society*, 83, 85. Francis
Tschan, *Saint Bernward of Hildesheim*, vol. 1, 50, traces his "ruinous faults"
to his mother. Tammo is described by Peter Damian in his "Life of Saint
Romuald," in Gerd Althoff, *Otto III*, 143. Eleanor Duckett describes
Otto's youth in *Death and Life in the Tenth Century*, 108–110.

201 *Gregory V:* Pierre Riché, *Gerbert d'Aurillac*, 165; Teta E. Moehs, *Gregorius
V, 996–999*, 25.

201 *mantle:* From the *Annals of Quedlinburg*; see Pierre Riché, *Les grandeurs de
l'an mil*, 243. For the illuminations, see Dominique Alibert, "Majesté ottonienne: L'hommage des nations à l'empereur," in Olivier Guyotjeannin
and Emmanuel Poulle, eds., *Autour de Gerbert d'Aurillac*, 82–87; Althoff,
cover of *Otto III*. For his letter, see Gerbert, 271.

202 *Rome:* See Althoff, 58–60; Anna Celli-Fraentzel, "Contemporary Reports
on the Mediaeval Roman Climate"; Jean Chelini, "Rome et le Latran au
temps de Sylvestre II," in Pierre Riché and Paul Poupard, eds., *Gerbert:
Moine, évêque, et pape*, 213–233; Paul Hetherington, *Medieval Rome*, esp.
3, 33, 42; and Riché, *Gerbert d'Aurillac*, 165–166, and *Les grandeurs*, 264–
267, 280–281.

204 *power struggles:* Moehs, 3, 37–42. Otto was aware of the Byzantine concept
of an emperor as the divinely appointed head of church and state; see Jenkins, *Byzantium*, 259; Adso of Montier-en-Der, "Book of the Antichrist,"
translated by Bernard McGinn, in *Apocalyptic Spirituality*, 85; and Norwich, 2–3.

204 *Adalbert:* Duckett, 113–115; Althoff, 65, 138–140; Moehs, 35; and Phyllis G. Jestice, "A New Fashion in Imitating Christ," in Frassetto, 165–185.

205 *"vehement grief":* Gerbert, 271.

205 *Robert:* Riché, *Gerbert d'Aurillac,* 168; Jim Bradbury, *The Capetians,* 12, 83; and Elizabeth M. Hallam, *Capetian France,* 70. As Constance Brittain Bouchard shows, the Church was revising the concept of "incestuous"; under the old rules, Bertha and Robert could marry, but under the new ones, there was no woman of sufficient rank in Europe whom Robert could legally wed. See *"Those of My Blood": Constructing Noble Families in Medieval Francia,* 40–46.

206 *Abbo set off again:* Abbo's letters are translated by Duckett, 130. Thomas Head cites Aimo of Fleury on Abbo being "more powerful" than the king in *Hagiography and the Cult of Saints,* 244.

207 *Gerbert did not wait:* For his letters about leaving Reims, see Gerbert, 280, 281, 221, 272, 282. The lavish copy of Boethius's *On Arithmetic* is discussed by Henri Focillon, *The Year 1000,* 157; Florentine Mutherich, "The Library of Otto III," in Peter Ganz, ed., *The Role of the Book in Medieval Culture,* 20.

209 *"magnificent German estate called Sasbach":* For his letters to Otto III, see Gerbert, 290, 287, 293, 292.

211 *"Saxon ignorance":* Letter from Otto III is from Gerbert, 294. For other translations, see Focillon, 157; C. Stephen Jaeger, *The Envy of Angels,* 56. Althoff attributes the idea that Otto was "un-German" to previous historians, 68.

211 *"ours is the Roman Empire":* Gerbert, 297, 296. Althoff notes Gerbert's appointment to the court chapel as "musician," 69; see also Riché, *Gerbert d'Aurillac,* 182. For Gerbert's treatise on reason, see Dominique Poirel, "L'art de la logique: Le *De rationali et ratione uti* de Gerbert," in Guyot-jeannin and Poulle, *Autour de Gerbert,* 312–320 (emphasis added). Mayr-Harting discusses arithmetic as a way to "tap into that cosmic whole which the ruler needed to grasp," 143, 165–166.

Chapter XII: The Pope of the Year 1000

213 *"minds of angry persons":* Gerbert, 90.

213 *John Philagathos:* Thietmar of Merseburg, 172–174; translator David Warner cites the *Annals of Quedlinburg* and Johannes diaconis in his notes. Ralph the Bald also records the affair, 25. Teta E. Moehs analyzes the sources, including the *Annals of Hildesheim* and the *Vita* of Saint Nil, in *Gregorius V,* 18, 55–66; as do Gerd Althoff, *Otto III,* 73–79; Eleanor Duckett, *Death and Life in the Tenth Century,* 124–127; and Pierre Riché, *Gerbert d'Aurillac,* 192–193. Francis Tschan discusses Philagathos as Otto

III's tutor, *Saint Bernward of Hildesheim*, vol. 1, 48–49. For Leo of Synada's letters, see Leo of Synada, "Ambassade de Léon de Synades, envoyé de Basile II à Otton III et au pape," in *Epistoliers byzantins du Xe siècle*, translated (into French) by J. Darrouzes; Leo of Synada, *The Correspondence of Leo, Metropolitan of Synada and Syncellus*, edited and translated by Martha Pollard Vinson.

217   *the papacy:* See Kathleen G. Cushing, *Reform and the Papacy in the Eleventh Century*, 17–23, 55–58, 82–85; G. Tellenbach, *The Church in Western Europe from the Tenth to the Early Twelfth Century*, 74. Moehs examines the cooperation between Otto and Gerbert, 80–86, as does Riché, *Gerbert d'Aurillac*, 201–221. Like Otto, Gerbert seems to have understood the Byzantine concept of the emperor as "the elect of God, the embodiment of divine and universal Providence"—and so the pope's overlord. See Romilly Jenkins, *Byzantium*, 259; Adso of Montier-en-Der, "Book of the Antichrist," translated by Bernard McGinn, *Apocalyptic Spirituality*, 85; and John Julius Norwich, *Byzantium*, 2.

219   *official papal documents:* Gerbert, 305–371; Lattin also publishes lists of Gerbert's writings that are "not extant," but attested to by previous historians, and "spurious," 381–389. For letters quoted, see Gerbert, 316, 326, 333, 356, 313, 324 (emphasis added); Lattin defines "indiction years" on 306n. The story of Sophie's long quarrel with Bernward of Hildesheim is told by Roland Allen, "Gerbert, Pope Silvester II," 659; Pierre Riché, *Les grandeurs de l'an mil*, 159; and Tschan (from Bernward's side), vol. 1, 162–195. Tschan discusses Bernward's artwork in volumes 2 and 3.

224   *Terrors of the Year 1000:* The history of the idea is discussed in Michael Frassetto, ed., *The Year 1000*; Richard Landes, Andrew Gox, and David C. Van Meter, eds., *The Apocalyptic Year 1000*. Jane Schulenburg dates the "clear signs" charters in "Early Medieval Women, Prophecy, and Millennial Expectations," in Frassetto, 246. See also Henri Focillon, *The Year 1000*, 70, on the "timeless" quality of the Apocalypse.

225   *Adso:* Adso of Montier-en-Der, 89–96. Otto's chancellor and friend, Herbert of Koln, owned a copy. See also Focillon, 57; Benjamin Arnold, "Eschatological Imagination and the Program of Roman Imperial and Ecclesiastical Renewal at the End of the Tenth Century," in Landes et al., 272–276; and Daniel Verhelst, "Adso of Montier-en-Der and the Fear of the Year 1000," in Landes et al., 82–87.

226   *Gniezno:* Thietmar of Merseburg, 182; for "the Anonymous Gaul," see Warner's notes, 184–185.

226   *Charlemagne's tomb:* Althoff, 104; Duckett, 133; Matthew Gabriele, "Otto III, Charlemagne, and Pentecost A.D. 1000," in Frassetto, 111–123; Thietmar of Merseburg, 185. For Gerbert's reaction, see Riché, *Gerbert d'Aurillac*, 218.

228 *"frenzied enemies"*: Gerbert, 350. On the revolts, see Althoff, 119–129; Duckett, 133–136; Moehs, 87–88; and Riché, *Gerbert d'Aurillac*, 222–227. "Are you not my Romans?" is recorded in *The Life of Saint Bernward*, translated here from Riché in *Gerbert d'Aurillac*, 222. Otto's penance in Ravenna is reported by Brun of Querfurt, as quoted by Benjamin Arnold, "Eschatological Imagination and the Program of Roman Imperial and Ecclesiastical Renewal at the End of the Tenth Century," in Landes et al., 279–280; Riché quotes a version recorded by Peter Damian in *Gerbert d'Aurillac*, 227.

Chapter XIII: The End of the World

231 *Princess Zoe:* See John Julius Norwich, *Byzantium*, 259, 269–310. Thietmar of Merseburg records Otto's death and succession, 187–189; the idea that his corpse was tied to a horse is an "embroidery" of later sources, says Gerd Althoff, *Otto III*, 129.

232 *"the black arts":* William of Malmesbury, 279–289. David Rollo explains the literary allusion in *Glamorous Sorcery*, 3–20.

235 *Gerbert died:* Roland Allen, "Gerbert, Pope Silvester II," 661, 664–668; Robert Favreau, "Le souvenir officiel: L'épitaphe de Silvestre II à Saint-Jean-de-Latran," in Olivier Guyotjeannin and Emmanuel Poulle, eds., *Autour de Gerbert d'Aurillac*, 336–341; and Pierre Riché, *Gerbert d'Aurillac*, 232–233. Both Allen and Riché quote Canon Caesar Raspo's account of the opening of Gerbert's tomb in 1648.

236 *"white mantle of churches":* Ralph the Bald, 115. The discovery of the second head of John the Baptist is told by Richard Landes in *Relics, Apocalypse, and the Deceits of History*, 47; he also details Ademar's forgeries and the riots over Saint Martial's relics, 269–279.

237 *"relic jamborees":* Landes, 47; he also cites Ademar of Chabanne's report of the collective pilgrimage to Jerusalem, 155. R. I. Moore, *Formation of a Persecuting Society*, points out the change from *miserere nobis* to *dona nobis pacem*, 320.

238 *heretics:* Ralph the Bald, 139–151; Ademar of Chabannes in Landes, 128. A monk from Ripoll who was present at the burning, Landes adds, "wrote his abbot soon thereafter with the news," according to the *Life of Gauzlin*. For the heretics' connection to Gerbert, see Moore, 12–15. On "The Condemnation of Aristotle's Books on Natural Philosophy in 1210 at Paris," see Edward Grant, *A Source Book in Medieval Sciences*, 42–43.

238 *religious intolerance:* Ralph the Bald blames the destruction of the Holy Sepulchre on the Jews, 133–137, as does Ademar of Chabannes in Landes, 40. Moore quotes Guibert of Nogent's description of the massacres of Jews, 28, and the general change in attitude toward Jewish citizens, 28–30,

76–83. The rise of intolerance in Spain is chronicled by Maria Rosa Menocal, *The Ornament of the World*, 96–103. She writes, "The old Andalusian order, with its political unity and cultural grandeur, exploded like a star," 100. On the importance of the *filioque* in dividing East and West, see John Man, *Atlas of the Year 1000*, 35; Norwich finds additional reasons, 315–322.

240 *Gerbert's reputation:* "great genius" is from Richer of Saint-Remy (before 998), vol. 2, 51. Helgaud of Fleury (c. 1041) and the son of Count Borrell are quoted by Riché, *Gerbert d'Aurillac*, 202, 234. Sergius IV (c. 1009–1012) is translated by Anna Marie Flusche, *The Life and Legend of Gerbert of Aurillac*, 75. "Surpassed his contemporaries," is from Thietmar of Merseburg (1013–1018), 303. "Acutely intelligent" is from Ralph the Bald (writing before 1030), 27. "Happy the state" is from Sigebert of Gembloux (c. 1070–1100), quoted by Henry Mayr-Harting, *Church and Cosmos in Early Ottonian Germany*, 60. "Among those shining, shone exceedingly" is quoted by Landes, 312.

241 *Beno:* Allen traces the "dark legend of Gerbert" to Benno of Osnabruck and imputes to it a political reason, 664–668; as does Riché, *Gerbert d'Aurillac*, 10; see also Darlington, "Gerbert the Teacher," 462. I follow Flusche in spelling his name "Beno"; rather than the German Benno from Osnabruck, the author of the dark legend was more likely the Italian Beno, cardinal-priest of the Church of SS Martino e Silvestro in Rome. The two may also be the same person; see Flusche, 82–86. The first to record the "R to R to R" joke was Helgaud of Fleury in his *Life of King Robert*; see Riché, *Gerbert d'Aurillac*, 234.

241 *Meridiana:* Walter Map, *De nugis curialium*, edited and translated by M. R. James, C. N. L. Brooke, and R. A. B. Mynors, 351–365.

242 *Saracen philosopher:* William of Malmesbury, 279–289.

243 *talking head:* For Ibn Juljul's description, see Julio Samso, "Astrology, Pre-Islamic Spain, and the conquest of al-Andalus," in *Islamic Astronomy and Medieval Spain*, Chap. 2, 86.

243 *"best necromancer":* Michael Scot (c. 1175–1234), in Lynn Thorndike, *Michael Scot*, 93–94.

243 *Martin the Pole:* Pascale Bourgain, "Silvestre II dans le *Liber pontificalis*," in Guyotjeannin and Poulle, 354–357.

244 *Luther's disciples:* Riché mentions the use of Gerbert's writing by the Protestants in his introduction to Gerbert's *Correspondance*, edited and translated (into French) by P. Riché and J.-P. Callu, xiv–xv; Riché discusses Cardinal Baronius in the same work, xvii–xix. See also Claude Grimmer, "Gerbert vu par le carme Géraud Vigier," in Guyotjeannin and Poulle, 358–360. Jules Michelet is quoted by Riché, *Gerbert d'Aurillac*, 13, and *Les grandeurs de l'an mil*, 14.

247 *"ancient chronicle":* Richard Erdoes, A.D. *1000: Living on the Brink of Apocalypse,* 1, 8.

247 *rediscover the Scientist Pope:* See Tim Roufs, "Inter-facing the Inevitable: Appropriate Technology in the 21st Century"; "Y1K: Web-Only Supplement to *The Millennium Issue,*" *Searcher Magazine,* http://www.infotoday.com/searcher/jan00/y1k.htm; Jay Ryan, "Gerbert d'Aurillac: Y1K's Science Guy," *Sky & Telescope* (February 2000): 38. See Pope John Paul II's letter at http://www.vatican.va/holy_father/john_paul_ii/speeches/1999/april. Also Keith Devlin's "The Mathematical Legacy of Islam," *MAA Online,* http://www.maa.org/devlin/devlin_0708_02.html; Kalpana Sahni's editorial from the Pakistan *Daily Times,* October 21, 2006, http://dailytimes.com.pk/default.asp?page=2006%5C10%5C21%5Cstory_21–10–2006_pg3_3; Pope Benedict XVI's message on the Year of Astronomy, http://www.vatican.va/holy_father/benedict_xvi/angelus/2008/documents/hf_ben-xvi_ang_20081221_en.html.

BIBLIOGRAPHY

Medieval Sources

Abbo of Fleury and Ramsey. *Commentary on the Calculus of Victorius of Aquitaine.* Ed. A. M. Peden. Oxford University Press, 2003.

Adalbero (Ascelin) of Laon. *Poème au roi Robert.* Ed. and tr. (into French) Claude Carozzi. Les Belles Lettres, 1979.

Ademar of Chabannes. *Chronique.* Ed. J. Chavanon. Collection de textes pour servir à l'étude et à l'enseignement de l'histoire 20, 1897.

Adso of Montier-en-Der. "Book of the Antichrist." In *Apocalyptic Spirituality: Treatises and Letters of Lactantius, Adso of Montier-en-Der, Joachim of Fiore, the Spiritual Franciscans, Savonarola.* Tr. Bernard McGinn. Classics of Western Spirituality. Paulist Press, 1979.

Aelfric of Eynsham. *AElfric's Colloquy.* Tr. G. N. Garmonsway. University of Exeter Press, 1978.

Anon. *Le Calendrier de Cordoue.* Ed. R. Dozy. Tr. (into French) Ch. Pellat. Brill, 1961.

Augustine of Hippo. *The City of God.* Tr. Marcus Dods. Christian Classics Ethereal Library at Calvin College. Http://www.ccel.org/ccel/schaff/npnf102.iv.XVI.9.html.

———. *The Literal Meaning of Genesis.* Tr. J. H. Taylor. Newman Press, 1982.

Benedict of Aniane. *Saint Benedict's Rule for Monasteries.* Tr. Leonard J. Doyle. Order of Saint Benedict, 1948; reprint, 2001. Http://www.osb.org/rb/text/toc.html#toc.

Bernard of Angers. "Liber miraculorum sancte Fidis." In *The Book of Sainte Foy.* Tr. Pamela Sheinborn. University of Pennsylvania Press, 1995.

Bernelin. *Libre d'abaque.* Ed. Béatrice Bakhouche and Jean Cassinet. Princi Néguer, 1999.

Boethius. *The Consolation of Philosophy.* Tr. E. Watts. Penguin Books, 1969.

———. *De institutione arithmetica.* In *Boethian Number Theory.* Ed. and tr. Michael Masi. Rodopi, 1983.

————. *De institutione musica.* Selections in *Source Readings in Music History.* Ed. O. Strunk. Norton, 1950.

Byrhtferth of Ramsey. *Byrhtferth's Manual,* A.D. *1011.* Ed. and tr. S. J. Crawford. Early English Text Society, 1929.

Fulbert of Chartres. *The Letters and Poems of Fulbert of Chartres.* Tr. Frederick Behrends. Clarendon Press, 1976.

Gerbert of Aurillac. *Correspondance.* Ed. and tr. (into French) P. Riché and J.-P. Callu. Les Belles Lettres, 1993.

————. *The Letters of Gerbert, with His Papal Privileges as Sylvester II.* Tr. Harriet Pratt Lattin. Columbia University Press, 1961.

————. *Opera mathematica (972–1003).* Ed. Nicolai Bubnov. Berolini, 1899.

Guibert of Nogent. *A Monk's Confession.* Tr. Paul J. Archambault. Pennsylvania State University Press, 1996.

Hasdai ibn Shaprut. "The Epistle of R. Chisdai, Son of Isaac (of Blessed Memory) to the King of the Khozars (ca. 960)." In *Jewish Travellers.* Tr. Elkan Nathan Adler. Routledge, 1930.

Hrotsvit of Gandersheim. *Hrotsvit of Gandersheim: A Florilegium of Her Works.* Tr. Katharina Wilson. D. S. Brewer, 1998.

Ibn Hauqal, Muhammad. *Configuration de la terre (Kitab surat al-Ard).* Tr. (into French) J. H. Kramers and G. Wiet. Commission internationale pour la traduction des chefs-d'oeuvre, 1964.

Ibn Juljul. "On the Translation of Dioscorides." In "The Early Growth of the Secular Sciences in Andalusia." Tr. George Hourani. *Studia Islamica* 32 (1970): 155–156.

Isidore of Seville. *Etymologies, Book II.* Tr. Peter K. Marshall. Les Belles Lettres, 1983.

Jean de Saint-Arnoul. *La Vie de Jean, Abbé de Gorze.* Ed. and tr. (into French) Michel Parisse. Picard, 1999.

John of Salerno. "The Life of Saint Odo of Cluny." In *St Odo of Cluny.* Tr. Gerard Sitwell. Sheed and Ward, 1958.

Leo of Synada. "Ambassade de Léon de Synades, envoyé de Basile II à Otton III et au pape." In *Epistoliers byzantins du Xe siècle.* Tr. (into French) J. Darrouzes. Archives de l'Orient chrétien, 1960.

————. *The Correspondence of Leo, Metropolitan of Synada and Syncellus.* Ed. and tr. Martha Pollard Vinson. Dumbarton Oaks Research Library and Collection, 1985.

Liudprand of Cremona. *The Works of Liudprand of Cremona.* Tr. F. A. Wright. Routledge, 1930.

Odilo of Cluny. "The Epitaph of Adelheid." In *Queenship and Sanctity.* Tr. Sean Gilsdorf. Catholic University of America Press, 2004.

Odo of Cluny. "The Life of Saint Gerald of Aurillac." In *St Odo of Cluny.* Tr. Gerard Sitwell. Sheed and Ward, 1958.

Radulf of Liege and Ragimbold of Koln. "Letters." In *Mémoires scientifiques*. Ed. Paul Tannery. Vol. 5, 229–303. E. Privat, 1887–1921.

Ralph the Bald (Rodulfus Glaber). *The Five Books of the Histories*. Tr. John France, Neithard Bulst, and Paul Reynolds. Clarendon Press, 1989.

Richer of Saint-Remy. *Histoire de France (888–995)*. Ed. and tr. (into French) Robert Latouche. H. Champion, 1930–1937.

Synesius of Cyrene. "On the Astrolabe." In *Livius.org: A Website on Ancient History*. Tr. A. Fitzgerald. Http://www.livius.org.

Thietmar of Merseberg. "Chronicon." In *Ottonian Germany*. Tr. David Warner. Manchester University Press, 2001.

Walter Map. *De nugis curialium*. Ed. and tr. M. R. James, C. N. L. Brooke, and R. A. B. Mynors. Clarendon Press, 1983.

William of Malmesbury. *Gesta Regum Anglorum*. Tr. R. A. B. Mynors, R. M. Thomson, and M. Winterbottom. Vols. 1 and 2. Clarendon Press, 1998.

Modern Sources

Allen, Roland. "Gerbert, Pope Silvester II." *English Historical Review* 7 (1892): 625–668.

Althoff, Gerd. *Otto III*. Pennsylvania State University Press, 2003.

Barthelemy, Dominique. *L'an mil et la paix de Dieu*. Fayard, 1999.

Bedos-Rezak, Brigitte. "Review of *Autour de Gerbert d'Aurillac*." *Speculum* 73 (1998): 528–531.

Bekken, Otto B. "Algorismus of *Hauksbók*." Http://home.hia.no/~ottob/.

Berggren, J. L. *Episodes in the Mathematics of Medieval Islam*. Springer-Verlag, 1986.

Berkhofer, Robert F. *Day of Reckoning*. University of Pennsylvania Press, 2004.

———, ed. *The Experience of Power in Medieval Europe*. Ashgate, 2005.

Bischoff, Bernhard. *Manuscripts and Libraries in the Age of Charlemagne*. Cambridge University Press, 1994.

Bonnery, André. *L'abbaye Saint-Michel de Cuixà*. MSM, 2005.

Borrelli, Arianna. *Aspects of the Astrolabe: "Architectonica Ratio" in Tenth- and Eleventh-Century Europe*. Franz Steiner Verlag, 2008.

Borst, Arno. *The Ordering of Time*. University of Chicago Press, 1993.

Bouchard, Constance Brittain. *"Those of My Blood": Constructing Noble Families in Medieval Francia*. University of Pennsylvania Press, 2001.

Bowman, Jeffrey. *Shifting Landmarks*. Cornell University Press, 2004.

Bradbury, Jim. *The Capetians*. Hambledon Continuum, 2007.

Brockett, Clyde W. "The Frontispiece of Paris, Bibl. Nat. Ms. Lat. 776: Gerbert's Acrostic Pattern Poems." *Manuscripta* 39 (1995): 3–25.

Brownrigg, Linda L., ed. *Making the Medieval Book*. Anderson-Lovelace, 1995.

Bruce, Scott G. *Silence and Sign Language in Medieval Monasticism.* Cambridge University Press, 2007.

Brunterc'h, Jean-Pierre. "Gerbert, archevêque de Ravenne et pape en Italie." In *Gerbert: Moine, évêque, et pape*, 225–268. Association catalienne pour la commémoration du pape Gerbert, 2000.

Bukofzer, F. "Speculative Thinking in Mediaeval Music." *Speculum* 17 (1942): 178–180.

Burnett, Charles. "The Abacus at Echternach in ca. 1000 AD." *SCIAMVS* 3 (2002): 91–108.

———. "Algorismi vel helcep decentior est diligentia: The Arithmetic of Adelard of Bath and His Circle." In *Mathematische Probleme im Mittelalter*, ed. Menso Folkerts, 221–331. Harrassotiz, 1996.

———. "King Ptolemy and Alchandreus the Philosopher." *Annals of Science* 55 (1998): 329–368.

———. *Magic and Divination in the Middle Ages.* Ashgate, 1996.

———, ed. *The Second Sense.* Warburg Institute, 1991.

Celli-Fraentzel, Anna. "Contemporary Reports on the Mediaeval Roman Climate." *Speculum* 7 (1932): 96–106.

Charbonnel, Nicole, and Jean-Eric Iung, eds. *Gerbert l'Européen: Acts du colloque d'Aurillac 4–7 juin 1996.* Société des lettres, sciences et arts "La Haute-Auvergnes," 1997.

Chelini, Jean. "Rome et le Latran au temps de Sylvestre II." In *Gerbert: Moine, évêque, et pape*, ed. Pierre Riché and Paul Poupard, 213–233. Association catalienne pour la commémoration du pape Gerbert, 2000.

Clemens, Raymond, and Timothy Graham. *Introduction to Manuscript Studies.* Cornell University Press, 2007.

Constable, Giles. "Dictators and Diplomats in the Eleventh and Twelfth Centuries." *Dumbarton Oaks Papers* 46 (1992): 37–46.

Cushing, Kathleen G. *Reform and the Papacy in the Eleventh Century.* Manchester University Press, 2005.

Darlington, Oscar G. "Gerbert, 'obscuro loco natus.'" *Speculum* 11 (1936): 509–520.

———. "Gerbert the Teacher." *American Historical Review* 52 (1947): 456–476.

Davids, Adelbert, ed. *The Empress Theophano.* Cambridge University Press, 1995.

Destombes, Marcel. "Un astrolabe carolingien et l'origine de nos chiffres arabes." *Archives internationales d'histoire des sciences* 15 (1962): 3–45.

Duckett, Eleanor. *Death and Life in the Tenth Century.* University of Michigan Press, 1967.

Eastwood, Bruce. "Calcidius's Commentary on Plato's *Timaeus* in Latin Astronomy of the Ninth to Eleventh Centuries." In *Between Demonstration and Imagination*, ed. A. Nauta and A. Vanderjagt, 171–209. Brill, 1999.

————. "Invention and Reform in Latin Planetary Astronomy." In *Latin Culture in the Eleventh Century: Proceedings of the Third International Conference on Medieval Latin Studies,* ed. Michael W. Herren, C. J. McDonough, and Ross G. Arthur, 264–291. Brepols, 2003.

————. *Ordering the Heavens.* Brill, 2007.

————. *Planetary Diagrams for Roman Astronomy in Medieval Europe.* American Philosophical Society, 2004.

————. *The Revival of Planetary Astronomy in Carolingian and Post-Carolingian Europe.* Ashgate, 2002.

Erdoes, Richard. A.D. *1000: Living on the Brink of Apocalypse.* HarperCollins, 1988.

Evans, Gillian R. "*Difficillima et Ardua:* Theory and Practice in Treatises on the Abacus, 950–1150." *Journal of Medieval History* 3 (1977): 21–38.

————. "Schools and Scholars: The Study of the Abacus in English Schools c. 980–c. 1150." *English Historical Review* 94 (1979): 71–85.

Evans, Joan. *Monastic Life at Cluny.* Oxford University Press, 1931.

Fletcher, Richard. *The Quest for El Cid.* Oxford University Press, 1989.

Flusche, Anna Marie. *The Life and Legend of Gerbert of Aurillac.* Edwin Mellen Press, 2006.

Focillon, Henri. *The Year 1000.* Harper, 1971.

Folkerts, Menso. *The Development of Mathematics in Medieval Europe.* Ashgate, 2006.

————. *Essays on Early Medieval Mathematics.* Ashgate, 2003.

————. "The Names and Forms of the Numerals on the Abacus in the Gerbert Tradition." In *Gerberto d'Aurillac: Da abate di Bobbio a Papa dell'anno 1000,* ed. Flavio G. Nuvolone, 245–265. Archivum Bobbiense Studii IV. Associazione culturale amici di Archivum Bobiense, 2001.

————. "Review of *De Verhandeling over de Cirkelkwadratuur van Franco van Luik van Omstreeks 1050* by A. J. E. M. Smeur." *Isis* 63 (1972): 272–274.

Fournier, Sylvie. *A Brief History of Parchment and Illumination.* Les éditions Fragile, 1998.

Frassetto, Michael, ed. *The Year 1000: Religious and Social Response to the Turning of the First Millennium.* Palgrave/Macmillan, 2002.

Freedman, Paul. *The Diocese of Vic.* Rutgers University Press, 1983.

Ganz, Peter, ed. *The Role of the Book in Medieval Culture.* Brepols, 1986.

Geary, Patrick. *Furta Sacra: Thefts of Relics in the Central Middle Ages.* Princeton University Press, 1978.

Gibson, Margaret, ed. *Boethius: His Life, Thought, and Influence.* Blackwell, 1981.

Glenn, Jason. *Politics and History in the Tenth Century: The Work and World of Richer of Reims.* Cambridge University Press, 2004.

Glick, Thomas. *From Muslim Fortress to Christian Castle.* Manchester University Press, 1995.

————. *Islamic and Christian Spain in the Early Middle Ages*. 2nd edition. Princeton University Press, 2005.

Glick, Thomas, Steven J. Livesey, and Faith Wallis, eds. *Medieval Science, Technology, and Medicine: An Encyclopedia*. Routledge, 2005.

Gottscher, Leandro. "Ancient Methods of Parchment-Making." In *Ancient and Medieval Book Materials and Techniques*, ed. Marilena Maniaci and Paola F. Munafo, vol. 1: 41–56. Biblioteca Apostolica Vaticana, 1993.

Grant, Edward. *A Source Book in Medieval Sciences*. Harvard University Press, 1974.

Grier, James. *The Musical World of a Medieval Monk: Ademar de Chabannes in Eleventh-Century Aquitaine*. Cambridge University Press, 2006.

Gunther, Robert T. *Astrolabes of the World*. Oxford University Press, 1932; reprint, 1976.

Guyotjeannin, Olivier, and Emmanuel Poulle, eds. *Autour de Gerbert d'Aurillac: Le pape de l'an mil*. Album de documents commentés: Matériaux pour l'histoire. L'école des Chartes 1, 1996.

Hallam, Elizabeth M. *Capetian France, 987–1328*. Longman, 1980.

Head, Thomas. "Art and Artifice in Ottonian Trier." *Gesta* 36 (1997): 65–82.

————. "The Development of the Peace of God in Aquitaine (970–1005)." *Speculum* 74 (1999): 656–686.

————. *Hagiography and the Cult of Saints*. Cambridge University Press, 1990.

————. "Letaldus of Micy and the Hagiographic Traditions of the Abbey of Nouaillé." *Analecta Bollandiana* 115 (1997): 253–267.

————. "Peace and Power in France Around the Year 1000." *Essays in Medieval Studies* 23 (2007): 1–17.

Head, T., and R. Landes, eds. *The Peace of God: Social Violence and Religious Response in France Around the Year 1000*. Cornell University Press, 1992.

Hentschel, Frank. "Gerbert, Organa, and Historical Thinking." In *Gerbertus qui et Silvester, minima gerbertiana da Piacenza a Lovanio, e altri studi a 1000 anni dalla morte del Pontefice (12. V. 1003)*. Archivum Bobiense 24 (2002): 53–77.

Hetherington, Paul. *Medieval Rome*. St. Martin's, 1994.

Hourani, George. "The Early Growth of the Secular Sciences in Andalusia." *Studia Islamica* 32 (1970): 155–156.

Ifrah, Georges. *From One to Zero: A Universal History of Numbers*. Viking, 1985.

————. *The Universal History of Numbers: From Prehistory to the Invention of the Computer*. John Wiley and Sons, 2000.

Iogna-Prat, Dominique. *Order and Exclusion: Cluny and Christendom Face Heresy, Judaism, and Islam (1000–1150)*. Cornell University Press, 2002.

Irving, Washington. *Columbus: His Life and Voyages*. G. P. Putnam's Sons, 1828; reprint, Read Books, 2007.

Jaeger, C. Stephen. *The Envy of Angels*. University of Pennsylvania Press, 1994.

————. *The Origins of Courtliness*. University of Pennsylvania Press, 1985.

Jenkins, Romilly. *Byzantium: The Imperial Centuries, AD 610–1071*. Weidenfeld and Nicolson, 1966.

Juste, David. *Les Alchandreana primitifs: Etude sur les plus anciens traités astrologiques latins d'origine arabe (Xe siècle)*. Brill, 2007.

———. "La sphère planétaire du ms. Vatican, BAV, Pal. lat. 1356 (XIIe siècle)." In *Entre Nadir et Zénith: Mélanges d'histoire des sciences offerts à Hossam Elkhadem*, ed. Jean-Marie Duvosquel, Robert Halleux, and David Juste, 205–221. Archives et Bibliothèques de Belgique, 2007.

Kennedy, E. S., and Marcel Destombes. "Introduction to *Kitab al'Amal bil Asturlab*." In *Studies in the Islamic Exact Sciences*, ed. David A. King and Mary Helen Kennedy, 405ff. American University of Beirut Press, 1983.

King, David. *Astrolabes and Angels*. Franz Steiner Verlag, 2007.

———. "Astronomical Instruments Between East and West." In *Kommunikation Zwischen Orient und Okzident Alltag und Sachkultur*, ed. Harry Kühnel, 143–198. Verlag der Österreichischen Akademie der Wissenschaften, 1994.

———. *The Ciphers of the Monks: A Forgotten Number-Notation of the Middle Ages*. Franz Steiner Verlag, 2001.

———. "The Earliest Known European Astrolabe in the Light of Other Early Astrolabes." *Physis* (1995): 359–404.

———. *In Synchrony with the Heavens*. Vols. 1 and 2. Brill, 2004–2005.

———. "Islamic Astronomy." In *Astronomy Before the Telescope*, ed. Christopher Walker, 143–174. St. Martin's Press, 1996.

Koetsier, Teun, and Luc Bergmans. *Mathematics and the Divine: A Historical Study*. Elsevier, 2005.

Koziol, Geoffrey. *Begging Pardon and Favor*. Cornell University Press, 1992.

Kunitzsch, Paul. "Al-Khwarizmi as a Source for the *Sententie astrolabii*." In *From Deferent to Equant*, ed. David King and George Saliba, 227–236. New York Academy of Sciences, 1987.

———. *Stars and Numbers*. Ashgate, 2004.

Landes, Richard. *Relics, Apocalypse, and the Deceits of History: Ademar of Chabannes, 989–1034*. Harvard University Press, 1995.

Landes, Richard, Andrew Gox, and David C. Van Meter, eds. *The Apocalyptic Year 1000*. Oxford University Press, 2003.

Lattin, Harriet Pratt. "Lupitus Barchinonensis." *Speculum* 7 (1932): 58–64.

———. "Origin of Our Present System of Notation According to the Theories of Nicholai Bubnov." *Isis* 29 (1933): 181–194.

———. "Review of *Assaig d'Historia de les Idees Fisiques i Matematiques a la Catalunya Medieval* by J. Millas Vallicrosa." *Speculum* 7 (1932): 436–438.

Lauranson-Rosaz, Christian. *L'Auvergne et ses Marges du VIIIe au XIe Siècle*. Cahiers de la Haute-Loire, 1987.

———. "De l'ancienne à la nouvelle noblesse." *Cahiers de la Haute-Loire* (2005): 185–198.

————. "*Theodosyanus nos instruit codex . . .* : Permanence et continuité du droit romain et de la romanité en Auvergne et dans le Midi de la Gaule durant le haut Moyen Âge." *Université Jean Moulin Cahiers du Centre d'Histoire Médié-vale* 3 (2005): 15–32.

Lawrence, C. H. *Medieval Monasticism.* 2nd ed. Longman, 1989.

Levy, Kenneth. *Gregorian Chant and the Carolingians.* Princeton University Press, 1998.

Leyser, Karl. *Communications and Power in Medieval Europe.* Hambledon Press, 1994.

————. *Rule and Conflict in an Early Medieval Society: Ottonian Saxony.* Illinois University Press, 1979.

Lindberg, David, ed. *Science in the Middle Ages.* University of Chicago Press, 1978.

Lorch, R. "The *Sphaera Solida* and Related Instruments." *Centaurus* 24 (1980): 153–161.

Lozovsky, Natalia. *"The Earth Is Our Book": Geographical Knowledge in the Latin West, ca. 400–1000.* University of Michigan Press, 2000.

MacKinney, Loren C. *Bishop Fulbert and Education at the School of Chartres.* Notre Dame, Medieval Institute, 1957.

Man, John. *Atlas of the Year 1000.* Harvard University Press, 1999.

Mayr-Harting, Henry. *Church and Cosmos in Early Ottonian Germany.* Oxford University Press, 2007.

McCluskey, Stephen C. *Astronomies and Cultures in Early Medieval Europe.* Cambridge University Press, 1998.

————. "Gregory of Tours, Monastic Timekeeping, and Early Christian Attitudes to Astronomy." *Isis* 81 (1990): 8–22.

McKeon, R. "Rhetoric in the Middle Ages." *Speculum* 17 (1942): 14.

McKitterick, Rosamund. *The Carolingians and the Written Word.* Cambridge University Press, 1989.

————. "Charles the Bald (823–877) and His Library: The Patronage of Learning." *English Historical Review* 95 (1980): 28–47.

Menocal, Maria Rosa. *The Ornament of the World: How Muslims, Jews, and Christians Created a Culture of Tolerance in Medieval Spain.* Little, Brown, 2002.

Miller, G. A. "The Formula ½a (a + 1) for the Area of an Equilateral Triangle." *American Mathematical Monthly* 28 (1921): 256–258.

————. "Gerbert's Letter to Adelbold." *School Science and Mathematics* 21 (1921): 649–653.

Moehs, Teta E. *Gregorius V, 996–999: A Biographical Study.* Hiersemann, 1972.

Moore, R. I. *Formation of a Persecuting Society.* Blackwell, 1987.

Murray, Alexander. *Reason and Society in the Middle Ages.* Clarendon Press, 1978.

Navari, Joseph V. "The Leitmotiv in the Mathematical Thought of Gerbert of Aurillac." *Journal of Medieval History* 1 (1975): 139–150.

Netz, Reviel, and William Noel. *The Archimedes Codex.* Da Capo Press, 2007.

Nightingale, John. "Oswald, Fleury, and Continental Reform." In *St Oswald of Worcester,* ed. Nicholas Brooks and Catherine Cubitt, 26–44. Leicester University Press, 1996.

Norberg, Dag. *Manuel pratique de latin médiéval.* Tr. R. H. Johnson. A. and J. Picard, 1968. Http://www.orbilat.com/Languages/Latin_Medieval/Dag_Norbert/index.html.

North, John. "Astrolabes." *Scientific American* 230 (1974): 196.

Norwich, John Julius. *Byzantium: The Apogee.* Knopf, 1992.

Nuvolone, Flavio G., "Appunti e novità sul *Carmen Figurato* di Gerberto d'Aurillac e la sua attività a Bobbio." *Archivum Bobiense* 25 (2003): 227–345.

———. "Elna e l'iscrizione attribuita a Gerberto d'Aurillac: Gerberto si recorda del vescovo e delle martiri?" *Archivum Bobiense* 29 (2007): 319–354.

———. "Gerbert d'Aurillac et la politique impériale ottonienne en 983: Une affaire de chiffres censurée par les moines?" In *Faire l'événement au Moyen Âge: Le temps de l'histoire,* ed. Claude Carozzi and Huguette Taviani-Carozzi, 235–261. Publications de l'Université de Provence, 2007.

———, ed. *Gerberto d'Aurillac: Da abate di Bobbio a Papa dell'anno 1000.* Archivum Bobiense Studii IV. Associazione culturale amici di Archivum Bobiense, 2001.

———, ed. *Gerberto d'Aurillac-Silvestro II: Linee per una sintesi. Atti del convegno internazionale: Bobbio, Auditorium di S. Chiara, 11 settembre 2004.* Archivum Bobbiense Studii V. Associazione culturale amici di Archivum Bobiense, 2005.

———. "Gerberto lacsia delle impronte: Iscrizione e monogramma. Ipotesi di lettura." *Archivum Bobiense* 27–28 (2005–2006): 257–320.

———, ed. *Gerbertus qui et Silvester, minima gerbertiana da Piacenza a Lovanio, e altri studi a 1000 anni dalla morte del Pontefice (12.V.1003).* Archivum Bobiense 24 (2002). Associazione Amici di Archivum Bobiense, 2003.

———. "La Presenza delle Cifre Indo-Arabe nel *Carmen Figurato* di Gerberto: Una discussione." *Archivum Bobiense* 26 (2004): 321–372.

Obrist, Barbara, ed. *Abbon de Fleury.* Publications universitaires Denis Diderot, 2006.

Pekonen, Osmo. "Gerbert d'Aurillac: Mathematician and Pope." *Mathematical Intelligencer* 22 (2000): 67–70.

Pijoan, Joseph. "Oliba de Ripoll (971–1046)." *Art Studies* 6 (1928).

Reston, James, Jr. *The Last Apocalypse.* Doubleday, 1998.

Riché, Pierre. *Gerbert d'Aurillac, le pape de l'an mil.* Fayard, 1987; reprint, 2006.

———. *Les grandeurs de l'an mil.* Bartilat, 1999.

Rollo, David. *Glamorous Sorcery: Magic and Literacy in the High Middle Ages.* Medieval Cultures 25. University of Minnesota Press, 2000.

Rosenwein, Barbara H., Thomas Head, and Sharon Farmer. "Monks and Their Enemies: A Comparative Approach." *Speculum* 66 (1991): 764–796.

Roufs, Tim. "Inter-facing the Inevitable: Appropriate Technology in the 21st Century." *13th Annual microCAD International Computer Science Conference, University of Miskolc, Hungary, 24 February 1999.* Http://www.d.umn.edu/cla/faculty/troufs/tr/Miskolc.html.

Russell, Jeffrey Burton. *Inventing the Flat Earth: Columbus and Modern Historians.* Praeger, 1991.

Samso, Julio. *Islamic Astronomy and Medieval Spain.* Ashgate, 1994.

———. "Review of David Juste, *Les Alchandreana primitifs.*" *Suhayl* 7 (2007): 183–187.

Savage-Smith, Emilie. *Islamicate Celestial Globes: Their History, Construction, and Use.* Smithsonian Institution Press, 1985.

Sigismondi, Costantino. "Gerberto e la misura delle canne d'organo." *Archivum Bobiense* 29 (2007): 355–396.

Stevens, Wesley M. "Compotistica et astronomica in the Fulda School." In *Saints, Scholars, and Heroes,* ed. Margot H. King and Wesley M. Stevens, vol. 2: 27–63. Hill Monastic Manuscript Library, 1979.

———. *Cycles of Time and Scientific Learning in Medieval Europe.* Ashgate, 1995.

———. "The Figure of the Earth in Isidore's *De natura rerum.*" *Isis* 71 (1980): 268–277.

Sullivan, Richard E., ed. *"The Gentle Voices of Teachers": Aspects of Learning in the Carolingian Age.* Ohio State University Press, 1995.

Tellenbach, G. *The Church in Western Europe from the Tenth to the Early Twelfth Century.* Cambridge University Press, 1993.

Thomson, Ron B. "Two Astronomical Tractates of Abbo of Fleury." In *The Light of Nature: Essays in the History and Philosophy of Science Presented to A. C. Crombie,* ed. J. D. North and J. J. Roche, 113–133. Springer, 1985.

Thorndike, Lynn. *Michael Scot.* Nelson, 1965.

Tosi, M., ed. *Gerberto. Scienza, storia e mito. Atti del Gerberti Symposium (Bobbio 25–27 Iuglio 1983).* Archivum Bobiense Studii II. Associazione culturale amici di Archivum Bobiense, 1985.

Tschan, Francis. *Saint Bernward of Hildesheim.* Vols. 1–3. University of Notre Dame, 1942–1951.

Warren, F. M. "Constantine of Fleury, 985–1014." *Transactions of the Connecticut Academy of Sciences* 15 (1909): 285–292.

Wolff, Philippe. *The Awakening of Europe.* Viking Penguin, 1968; reprint, 1985.

Zuccato, Marco. "Gerbert of Aurillac and a Tenth-Century Jewish Channel for the Transmission of Arabic Science to the West." *Speculum* 80 (2005): 742–763.

———. "Gerbert's Islamicate Celestial Globe." In *Gerberto d'Aurillac-Silvestro II: Linee per una sintesi. Atti del convegno internazionale: Bobbio, Auditorium di S. Chiara, 11 settembre 2004,* edited by Flavio G. Nuvolone, 167–188. Archivum Bobbiense Studii V. Associazione culturale amici di Archivum Bobiense, 2005.

INDEX

Aachen
  cathedral's processional cross, 200
    (fig.)
  as imperial city, 165, 173
  Otto III opens Charlemagne's tomb,
    226–228
  sacked by Lothar, 180
Abacus boards
  abacus as term, 82
  constructed by Gerbert, 5, 73, Plate
    5 (center section)
  described, 82–85, 88–89
  Gerbert's compared to Chinese,
    Roman, 81–82
  influence mathematics, algorithm
    system, 90
  introduce Arabic numbers, 79, 81
  manuscripts containing, 85 (fig.),
    88, Plate 5 (center section)
Abbasid Empire, 43, 45, 48
Abbo of Fleury (abbot), 100
  background, profile, 94–97
  compared to Gerbert, 97–98
  dislikes Constantine, 148–149
  disputes with Arnulf, defends
    Arnoul, 191–193
  and End Times rumors, 2, 100
  as Gerbert's enemy, 2, 94–95,
    206–207
  as monastery reformer, 192
  on papal authority, 197–198,
    223

on Year 1000, 100, 224, 245
  death of, 96
Abd al-Rahman (caliph of al-Andalus),
  43–44
Abd al-Rahman III (caliph of al-
  Andalus), 49
Achilleidos (Statius), 115
The Acts of Saint-Basle (Gerbert), 192,
  207, 244
Adalbero of Reims (archbishop), 211
  accused of treason, 142, 183, 185,
    211
  conspires against Lothar, 179,
    182–183
  and Egbert of Trier, 116
  forms coalition against Henry the
    Quarreler, 172–174
  Gerbert as friend, secretary,
    confidant, 66–69, 171, 188
  death affects Gerbert, 150, 153,
    187–188
  places Hugh Capet on throne, 179,
    186
Adalbert of Prague (bishop), 204–205
  martyred by Prussians, 208–209
  and Otto III's pilgrimage to tomb,
    226
Adalbold of Liege (bishop of Utrecht)
  discusses area of triangle with
    Gerbert, 3, 5, 46, 109–111, 110
    (fig.)
  on sphere's volume, 109

Adelaide (princess)
  eldest daughter of Otto II and
    Theophanu, 155
  abbess of Quedlinburg, 216, 223
Adelaide (empress)
  animosity toward Theophanu, 155,
    171, 196
  background and marriage to Otto I,
    161–162
  left in charge of Germany by Otto
    III, 199
  as regent for Otto III, 174
  relationship to Gerbert, 161, 196
Adelard of Bath, 90
Ademar of Chabannes, 52, 146, 227,
  237
Adso of Montier-en-der (abbot), 2, 30,
  225–226
Aelfric of Eynsham, 35
Aeneid (Virgil), 31, 36
Al-, *See under second syllable of personal
  names*
Al-Andalus region
  described, 42, 48–50, 54–55
  astrolabe and abacus sources, 52
  challenged by Guifre, 44
  and Hasdai ibn Shaprut, 51
  as home of math and science, 45
  and al-Khwarizmi, 134
  See also Catalonia, Islamic Spain
Alchandreana astrology texts,
  144–147
Alcuin of York, 35, 39, 93, 202
Alexander the Great, 146, 199
Algebra, 46, 246
Algorismus (al-Khwarizmi), 90–91
Algorithm, 46, 90
Almagest (Ptolemy), 118
Anaximander of Miletus, 124–125
Annals of Hildesheim, 226
Annals of Quedlinburg, 215
Annals of the Church (Baronius), 245

Anno Domini (A.D.) system, 99–100,
  224
Antichrist
  to be loosed in 1000, 2
  fables about Gerbert perpetuated,
    247
  as the pope, according to Arnulf,
    194–195
Antipodes debate, 128
Antipopes
  battle for precedence, 237
  Boniface VII, 195
  Clement III, 241
  John Philagathos (John XVI), 205,
    206, 213–215
  Leo VIII, 194
Apocalypse
  book of, 57
  dates recalculated, 100
  and increased intolerance, 236–238
  linked to Last Emperor, 224–225
  scenes in Otto III's robe, 201
  See also End of the World
The Arabian Nights (Scheherazade),
  45–46
Arabic language
  appearing in Latin manuscripts,
    134, 143–145
  as language of science, 4–5, 45
  as lingua franca in Spain, 51
Arabic numerals (nine-number
    system)
  brought to Spain from Baghdad, 89
  contained in Codex Vigilanus,
    58–59, 89
  explained by al-Khwarizmi, 46, 89
  found with abacus board in
    manuscripts, 80, 82, 85 (fig.), 88
  hidden in *Carmen Figuratum*, 167
    (fig.), 167–168
  introduced through Gerbert's
    abacus, 79, 88

not accepted for business
transactions, 89, 91
oldest manuscript containing, 89
used to teach math, 5, 88, 90–91
Western versions differ from
Baghdad's, 91
zero as place holder, 85
See also Indian numerals
Archimedes codex, 111–112
Arduin of Ivrea (margrave), 222
king of Italy, 232
Aristotle
on earth as a globe, 125
reading punishable by
excommunication, 238
as source for Gerbert's geometry,
108
works translated by Boethius, 38
Arithmetic, 39. *See also* Computus,
Mathematics
Armillary spheres, 5, 121–123, 136,
138
Arnoul of Reims (archbishop)
betrays King Hugh, 188–190
confesses guilt, 192–193
supported by John XV, Abbo,
195–196
unseated, replaced by Gerbert, 191
as Reims archbishop, 207
Arnulf of Orleans (bishop)
antipapal tirade, 193–195, 244
disputes with Abbo, 191–193
and naming of abbot of Fleury, 149
saves Hugh from fealty to Otto II,
180
to whom Gerbert entrusts property,
207
Ascelin of Laon (bishop)
accused of adultery with Queen
Emma, 173, 184–185
captured by Charles, escapes, 186,
187

codifies feudalism, 14
former student of Gerbert, 14, 90
names Gerbert as wizard, 145
tricks Arnoul, Charles, into being
captured, 190
Ascelin the German
gives astrolabe book to Constantine,
148, 149–150
Astrolabes
and Arabic astronomy, 47–48, 59,
133–134
constructing, 137–139
described, 5, 133–134, Plate 3
(center section)
Destombes astrolabe, 140–141 (fig.)
explained by al-Khwarizmi, 47–48
in first Latin book with Arabic
terms, 144
first reference in Latin, 139
Fulbert's use, instruction, 143–144
Gerbert's relation to, 134–135,
147–150
integrated into quadrivium, 143
treatise by Gerbert's student, 107
(fig.), 147–150
treatise by Maslama, 59
used by Otto III, 211–212
Astrology
as both science and fortune-telling,
59, 116
Gerbert's knowledge, 145–147
as lesson in divine harmony, 113
with *mathesis* referring to fortune-
telling, 59, 145
oldest Latin manuscript with Arabic,
144–145
revealed in *Alchandreana*, 144–147
Astronomy
and Arabic science, 45, 47–48, 59,
133–134
brought north from Cordoba, 52
al-Khwarizmi's influence, 47

Astronomy (*continued*)
  and models of the universe,
    121–123, Plate 6 (center section)
  and Ripoll 225, 57
  studied with spheres, 114–123
  as subject of quadrivium, 39
Ato of Vic (bishop)
  as mentor of Gerbert, 41, 44–45
  murdered by Narbonne archbishop,
    64, 75
  relationship to Borrell, 44–45
  travels with Gerbert to Rome,
    63–64
Augier of Reims, 189, 193
Aurillac, France, 12, 22–23, 249. *See
  also* Saint-Gerald's monastery
Ayrard, 30–31
Azalais of Anjou (countess),
    181–182

Baghdad, 4, 43
Bamberg Apocalypse, 224, 225 (fig.)
Barcelona
  Charlemagne battles Abd al-
    Rahman, 43
  sacked by al-Mansur, 42, 142, 223,
    239
Baronius (cardinal), 245–246
Basques, 43–44
Al-Battani, 118, 119
Beatrice of Lorraine (duchess), 184
Beatus of Liebana, 224
Bede. *See* Venerable Bede
Benedict V (pope), 194
Benedict VI (pope), 177, 195
Benedict VIII (pope), 217
Benedict XVI (pope), on Gerbert,
    248–249
Benedictines
  as makers of books, 25
  monastery life described, 15–18
  in monastery of Aurillac, 15–17

  in monastery of Fleury, 95–96
  rules (*see* Rule of Saint Benedict)
Beno (cardinal), 241, 245
Bernard of Angers, 21–22
Bernelin, 82, 88
Bernward of Hildesheim (bishop)
  in conflict over Gandersheim,
    220–222
  education of, 32
  evangelary of, 34 (fig.), Plate 2
    (center section)
  mediates for peace with Romans,
    228
  and schoolmaster Thangmar, 32
Bertha of Blois (countess), 206–207,
    217
Berthold of Breisgau (count), 215, 216
Al-Biruni, 47
Bishop of Rome, as term for papacy,
    196–197, 204, 217
Black magic
  fables created about Gerbert,
    134–135, 232–234, 241–243
  Gerbert's reputation, 79, 240
  linked to Gerbert's knowledge of
    astrology, 145–146
  stories discounted by later scholars,
    245–246
Bobbio monastery in Italy
  Gerbert as abbot, 158, 159–161
  Gerbert returns to Reims, 6, 68
  library, 23, 159
  stripped of its wealth, 160, 223
Boethius
  On Arithmetic, 101–103
  On Astrology, 30, 159
  On Music, 104–105
  Christianizes Pythagoras, 101–102
  as influential Aristotelian, 38
  quest for unity, 102–104
  as source for Gerbert's of geometry,
    108

Boleslav Chobry (duke of Poland), 219, 226

Bolingbroke, Henry St. John, 129

Boniface VII (antipope), 177, 195

Book buying as a profession, 49

Book-making
binding and cover process, 29–30
ink production, 28–29
parchment production, 25–28
rubrication and art work, 29

Book of Arithmetic Needed by Scribes and Merchants (al-Buzjani), 89

The Book of Miracles of Saint Foy (Bernard of Angers), 20, 21–22

Book of the Abacus (Bernelin), 82

Book of the Abacus (Fibonacci), 90

Book of the Antichrist (Adso), 225–226

Book of Wisdom, 39, 102

Book on the Abacus (Gerbert), 69–70, 73–74, 83, 89–90

Books
Gerbert as collector, 30–31
as keys to wisdom, 31
oldest Latin manuscript with Arabic words, 144–145
read in monasteries, 23–24
translation of, 4, 46, 51–52, 58, 59, 76, 101, 112, 118, 134, 236

Borrell of Barcelona (count), 41–42, 44–45
arqueta (reliquary) given by al-Hakam II, 52–53, Plate 4 (center section)
attends Ripoll cathedral consecration, 58
and kings of France, 142
and sacking of Barcelona by al-Mansur, 142–143
son, 232, 240
travels with Gerbert, Ato, to Rome in winter, 63–64, 75

Bruno (cousin of Otto III), 201. *See also* Gregory V

Bruno of Querfurt, 208–209

Al-Buzjani, Abul Wafa, 89

Byrtferth of Ramsey, 97

Byzantium/Byzantine Empire
and Archimedes codex, 112
influence in Ravenna, 154
Last Emperor legend, 200
and Otto II, 162–164
as source for royal brides, 186, 206, 214, 226, 240
and Theophanu, 6, 155

Caesar Augustus, 12, 200 (fig.), 233

Calcidius, 104, 108, 117, 123

Calendar of Cordoba (Racemundo), 49

Calculus, 111–112

Calendars
Anno Domini (A.D.) system, 99–100, 224
dating of Easter, 99
made with computus, 98

Calonimus the Jew, 162–163

Calvinists, 244

Canons, 65, 67–68

Capet, Hugh (king). *See* Hugh Capet

Capetian dynasty, 186–187

Carmen Figuratum (puzzle poem), 164 (fig.), 165–169, 167 (fig.), 211

Carolingian minuscule script, 33

Carolingians, 23, 74, 179, 186, 189

Castel Sant'Angelo in Rome, 177, 195, 203, 215, 229

Catalonia
astrolabe introduced, 134
church-state work in harmony, 44–45
Guifre encourages Visigoths to settle, 44
independence from France, 142–143

Catalonia (*continued*)
  legal justice system, 42–43
  nine-number system differs from
    Baghdad's, 91
  See also Al-Andalus, Islamic Spain
Cathedral schools
  become first universities, 68
  train noblemen, 65
  See also monastery schools
Catholicism. *See* Roman Catholic
    Church
Celestial Sphere, 113. *See also*
    Spheres/celestial globes
Charlemagne
  battles Abd al-Rahman, 43–44
  changes crowning date using A.D.
    system, 99–100
  line, dynasty, ends, 2, 6, 179
  revives puzzle poems, 165
  systemizes teaching in monasteries,
    33, 35
  tomb opened by Otto III, 226–228
Charles of Lorraine (duke)
  as last Carolingian, 74, 186
  influences King Louis V, 184
  pillages, kidnaps nobles, for
    kingship, 189–190
  pitted against Hugh Capet, 74,
    186–187
  as pretender to French throne,
    185–189
Chartres cathedral, 143, 147, 237
Chivalry code, 13
Christian empire as ideal. *See* Holy
    Roman Empire
Chronicon (Thietmar), 155
The Church. *See* Roman Catholic
    Church
Church and state in Catalonia, 44–45,
    218
Church of the Holy Sepulchre, 56, 239
Churches. *See* Roman Catholic Church

Cicero
  eloquence of, 127
  on friendship, love, 72
  as preferred author of Gerbert, 4,
    31, 36, 159
  on virtue for orators, 39
The City of God (Saint Augustine), 128
Civil war between Hugh and Charles
    of Lorraine, 189–190
Clamor, 192
Clement III (antipope), 241
Clement XI (pope), 219
Climate circles, 119–121, 136, 138,
    143, 178
Clocks
  astrolabes, 133, 137
  celestial, 120–121
  clepsydra, 136
  sundials, 114, 136
Cluny monastery, 18
Codex of Archimedes, 112
Codex production, 26–28
Codex Vigilanus, 58–59
Columbus, Christopher, 128–130, 247
Commentary on the Dream of Scipio
    (Macrobius), 127
Commentary on Victorius's Calculus
    (Abbo), 94, 99, 100–101
*The Compendious Book on Calculation
    by Completing and Balancing*
    (al-Khwarizmi), 46
Compostela in Spain, 19 (fig.), 22, 54,
    239
Computer, 5, 45, 46, 81, 137, 248
Computus, 97–98, 100, 127
Conques, France, 20–21
The Consolation of Philosophy
    (Boethius), 38
Constantine (Roman emperor), 62,
    165, 218, 224, 235
Constantine of Fleury
  as abbot of Micy, 149

as abbot of Nouaille, 149
and astrolabe book, 148, 149–150
on celestial sphere, 147
conflicts with Abbo, 95, 149
and Gerbert's *Book on the Abacus*,
  69–70, 83, 147
as Gerbert's student and close friend,
  69–70, 150
preserves Gerbert's letter collection,
  70–71 (fig.), 147–148, 244–245
Constellations
knowledge required to make
  astrolabes, 136
laid out from God's-eye view, 119,
  127
spheres used for instruction, 73,
  120–121
Cordoba, Spain
destroyed by fire, 239
intellectual life, 51–52
Royal Library, 4, 49–50
Cosmas Indicopleustes, 125–126, 131
Counting boards, 82, 85–86, Plate 5
  (center section). *See also* Abacus
  boards
Crescentian family
and Gregory of Tusculum, 232
and Pope John XV, 195, 199
replaces Pope Gregory with
  antipope, 205
strangles Pope Benedict VI, 177,
  195
supports antipope Boniface VII,
  177, 195
Crescentius III, 229, 232
Crescentius of the Marble Horse, 195,
  214–215
Crusades, 231, 238, 239
First Crusade in 1096, 7, 44
Cults
of saints, 21–22, 237–238
of the Virgin, 237

Cuxa monastery and cathedral
Arabic science influences design, 55
  (fig.), 56–57
with ascetic Romuald, 229
described, 22–23, 54–55, 56
under Abbot Garin, 56, 229

D'Angers, David, 249
Dark Ages
and End Times predictions, 2
of Gerbert replaced by intolerance,
  conflict, 236, 238
as term, 129
as time of science, enlightenment, 4,
  24, 84, 109–110
Dark Legend of Gerbert, 243–249
De astrologia (translated by Lobet), 58,
  144–145, 147
Debates
over Antipodes, 128
in classrooms for students of
  dialectic, 38–39, 84
in councils over archbishopric of
  Reims, 195–198
organized by Gerbert for kings, 5,
  208, 213
over Mercury and Venus, 122
over papal power vs. bishops'
  authority, 193–194
on physics as subdiscipline of
  mathematics, 123, 156–157, 204
*The Deeds of the Church of Rome vs.
  Hildebrand* (Beno), 241
Demosthenes, 23, 31
Dennis the Humble, 99–100
Description of the World (ibn Hauqal),
  48
Destombes astrolabe, 140–141 (fig.),
  143
Dhimma as tolerance creed, 50
Dialectic
Gerbert's teaching at Reims, 73

Dialectic (*continued*)
linked to Aristotle through Boethius, 37–39
Dietrich of Metz (bishop), 178
Dioscorides, 51, 243
Donatus, 35
Draper, John W., 131

Earth shaped as globe, 124–131, Plate 7 (center section)
circumference calculations, 5, 47, 125–127, 130
Irving's fiction, 129–131, 247
Easter date calculations, 98–99
Eastern Orthodox Church, 7, 39, 240
Echternach monastery, 79–81, 117
Eclipses, 120, 123–125, 133
Education
memorization techniques, 36–37, 86
Otto III's thirst for wisdom, 211–212
quadrivium disciplines, 39–40, 73, 79–81, 90, 93–94, 98–99, 113, 120, 122, 143, 168
and schoolmaster Raymond of Laval, 32
teaching of Latin, 34–36
trivium disciplines, 32, 33–39, 65, 73
writing, 33–34
Egbert of Liege, 35
Egbert of Trier (archbishop), 116, 172, 173–174, 182
Elne cathedral, 61
meaning of Gerbert's graffiti, 61–62, 211
Emma (queen of France)
accused of adultery with Ascelin, 173, 184–185
captured by Charles, 186

conspires against Hugh Capet, 181
daughter of Empress Adelaide, 173
marries Lothar, 173
uses Gerbert as secretary, 178, 187
Emperors. *See* Holy Roman Emperors
Empire. *See* Holy Roman Empire
Encyclopedia of Isidore of Seville, 42–43, 58, 60, 89
End of the World
and Charlemagne's inauguration date, 99–100
fables about Gerbert discounted, 247–248
linked to Last Emperor legend, 225–226
predictions, 1–2
Year 1000 fables persist, 224, 245–246
See also Apocalypse
End Times. *See* End of the World
Equinoxes
dating of, 98
day-length calculations, 114
precession, 139, 141
Eratosthenes, 112, 125–127
Euclid, 108
Excommunication
of Abbo, 96
of Gerbert, 196, 204, 205, 216
of King Robert, 194
sentence for reading Aristotle, 238
with shunning as punishment for monks, 17

Farfa monastery in Italy, 23
Feudalism and feudal code, 14, 220
Fibonacci, 90–91
Fields of Mars in Rome, 232–234, 245
Figeac monastery, 21
Finger counting or finger numbers, 86–87, 94, 98, 168
Firmicus, Julius, 145

Flat Earth Error and Flat Earthers,
125–126, 128–131
Fleury monastery
Abbo chosen as abbot instead of
Constantine, 178
as center of monastic reform, 96
described, 95
Flodoard of Reims, 74
Fortune-telling as aspect of astrology
Maslama of Madrid, 59
in old Latin manuscript with Arabic
words, 144–145
uses abacus, astrolabe, mathematics,
146
uses celestial spheres, 116
The Fourth Trumpet of the Apocalypse
manuscript, 235 (fig.)
Franco of Liege, 111
Friendship and love
Ciceronean code, 72
Gerbert's gift for, 150, 170
valued as sign of respect, 69–72, 201
Fulbert of Chartres (bishop)
peace hymn, 237
understands, explains, astrolabe,
143–144, 147
writes poems on scientific concepts,
37, 143
Fulda privilege, 223–224

Gabor, Szinte, 249
Galen, 46
Gandersheim convent and church,
220–222
Garin of Cuxa (abbot), 56–58
Gausbert of Mettlach, 80–81, 85 (fig.),
88, 116
Geometry
advanced textbook by Gerbert,
106–109, 107 (fig.), 110 (fig.)
as experimental science, 111, 139
as glimpse into mind of God, 108

pi explained in Archimedes codex,
112
See also Triangles
Geometry I (Boethius), 108, 109
Gerald of Aurillac (abbot)
correspondence with Gerbert, 56,
142, 171, 175
and Louis V, 142
permits Gerbert to leave Aurillac,
40–41
as youthful Gerbert's abbot, 14–15
Gerald the Good of Aurillac (count)
background, 11–13
founds monastery at Aurillac, 11
perfect facial image, 243
travels to Rome, 64
See also Saint Gerald of Aurillac
Gerann (schoolmaster of Reims), 65,
68, 105
Gerbert of Aurillac
background, personal and scientific,
4–6, 13–15
and abacus boards, 5, 79–91
as abbot of Bobbio, flees to Pavia,
158–161, 165–171
accused of necromancy, 145–146
accused of treason by Lothar, Louis
V, 183, 185
acts as pope, 219–223, 228–229
as Adalbero's confidant, Reims
canon, 65, 68
appointed pope by Otto III, 216
as archbishop of Ravenna, 1–3
as archbishop of Reims, 191,
205–206, 207–208
and astrolabes, 134–150
and Boethius' *Arithmetic,* 102–103,
210–211
as book collector, writer, scholar,
30–33, 112
brings mathematics, astronomy,
north from Spain, 52–53

Gerbert of Aurillac (*continued*)
  Carmen Figuratum (puzzle poem),
    164 (fig.), 165–169, 167 (fig.),
    211
  excommunicated by John XV, 196,
    204, 205
  friendship with Miro Bonfill,
    59–62
  and geometry, 106–111
  letter collections, 60, 70–71 (fig.),
    244–245
  the *Saltus Gerberti,* 103
  and spheres, 73, 114–121
  and treatise on organ pipes,
    103–106
  See also Sylvester II
Gerbert the Wizard, 241, 243, 246
Ghubar numbers, 88
Giant Bible of Echternach, 79–80
Godescalc of Le Puy (bishop), 19 (fig.),
  19–20, 22
Godfrey of Verdun (count)
  commissioned Gerbert to write
    letters, 178
  imprisoned by Lothar, 182–184
  released by King Hugh, 186
Gog and Magog
  defeated through baptism, 219
  descend on Germany, 199
  foretold in legend, 200
  as Vikings, Magyars, Saracens,
    Huns, 1, 219
Golden Age of Science, 240–241
Golden Age of Spain, 51
Gotmar of Girona (bishop), 53
Graffiti, sacred, 61, 62 (fig.), 211
Gran, Hungary, 219
Great Schism of 1054, 7, 231, 240
Greek classics
  and Petrarch, 129
  translated by Boethius, 38
Gregorian chants, 105

Gregory IV (pope), 207
Gregory V (pope)
  cousin of Otto III, 201
  excommunicates King Robert, 206,
    207
  Gerbert's puzzle epitaph, 219
  given name Bruno, 201
  and mutilation of Philagathos,
    215–216
  replaced by antipope through
    Crescentians, 205, 214
  rules on Adalbert, 204–205
  succeeded by Gerbert, 213
  See also Bruno
Gregory VI (pope), 68
Gregory VII (pope), 241. *See also*
  Hildebrand
Gregory the Great (pope), 23
Gregory of Tours, 114
Gregory of Tusculum, 228–229,
  232
Guibert of Nogent, 239
Guifre the Hairy (count), 44, 54
Guild of Money Changers, 91

Al-Hakam II (caliph)
  gives *arqueta* to Borrell, 52–53
  as scholar, book collecter, 49, 59
  death of, 142
Hand B scribe, 80
Harun al-Rashid. *See* Al-Rashid,
  Harun
Hasdai ibn Shaprut (vizier)
  describes al-Andalus, 48
  intellectual, politician, in Spain, 51,
    53
  as mathematician, 56
  and *On Medicine* translation, 51,
    243
Henry IV (emperor), 241
Henry of Bavaria (duke, king of
  Germany), 229, 230, 231–232

Henry the Quarreler (duke)
  aspires to regency of Otto III,
    171–174
  challenges Otto III, Theophanu, 6
Herbert of Koln, 230, 231
Heresy
  accusation as method to control
    churchmen, 238
  and round shape of earth, 124
Hildebrand (pope), 241. *See also*
    Gregory VII
Hipparchus of Rhodes, 136–137, 139
History of France (Michelet), 246, 249
History of France (Richer), 73–75, 79
History of Rome (Livy), 23
*The History of the Conflict Between
    Religion and Science* (Draper), 131
History of the English (Bede), 23
History of the Inductive Sciences
    (Whewell), 131
*History of the Warfare of Science with
    Theology* (White), 131
Holy Roman Emperors. *See* Henry IV,
    Otto I, Otto II, Otto III
Holy Roman Empire
  ideal imagined by Gerbert, Otto III,
    Adalbero, 6–7, 66, 199–200, 218,
    231
  and Petrarch's version of history, 129
  portrayed in *Carmen Figuratum*, 169
  ruled by regents Theophanu and
    Adelaide, 174
  shatters upon Otto III's death, 232
Horace, 23, 31, 36–37, 83
Horologium, 136, 208
Horoscopes, 47, 146–147. *See also*
    Astrology
House of memory technique, 37, 84
House of Wisdom, Baghdad
  of al-Rashid, al-Mamun, 45–46
  astrolabes used, 133–134
  and Dioscorides' *On Medicine*, 51

known for translations of classics, 4,
    46
observatory and astronomers, 47,
    133
science spreads to Christian Europe,
    7
Hraban Maur, 100, 166
Hrosvit of Gandersheim, 50
Hugh Capet (king of France)
  background, 179–182
  and Borrell's petition for aid,
    142–143
  crowning serves goal of empire,
    211
  dismisses Gerbert's, Adalbero's,
    treason charges, 185
  placed on throne by Adalbero,
    Gerbert, 6, 74, 186
  shields Gerbert, Adalbero, from
    Lothar, 183–184
  supports Gerbert as Reims
    archbishop, 195–196
  death of, 205
  See also Capetian dynasty
Hungary, 6, 219, 238, 248, 249
Hypatia, 137

Ibn Hauqal, Muhammad, 48
Ibn Ishaq, Hunayn, 46
Ibn Juljul, 51–52, 243
Ibn Qurra, Thabit, 46
Ibn Shaprut, Hasdai. *See* Hasdai ibn
    Shaprut
Imiza, 170, 171, 173
Indian numerals, 46, 58, 88–89. *See
    also* Arabic numerals (nine-
    number system)
Inquisition, 231, 238
Introduction to Grammar (Priscian), 35
Introduction to the Categories of Aristotle
    (Porphyry), 23, 38
Irving, Washington, 129–131, 247

Isidore of Seville (bishop)
  on computus, 98
  encyclopedia (Codex Vigilanus),
    42–43, 58, 60, 89
  On the Nature of Things, 124, 127
  on round shape of earth, 126
Islam
  and al-Andalus creed of tolerance,
    50
  as religion shared by Spanish
    Berbers, Muslims, 43
Islamic Spain
  as center of learning, 4
  Gerbert's studies in Barcelona,
    41–44
  origins of modern science, 248
  science spreads to Christian Europe,
    5, 248
  as tolerant society, 50–51
  uses al-Khwarizmi's science, 48

Jefferson, Thomas, 129
Jews
  knight assists Otto II, 162–163,
    239
  as People of the Book, 50
  scholars translate books from Arabic
    to Latin, 4, 52
  taxes paid, 223
  and tolerance in al-Andalus, 50–52
  persecuted, segregated, post-year
    1000, 239–240, 247
  See also Hasdai ibn Shaprut (vizier)
John XII (pope), 194
John XIII (pope), 64, 194–195
John XIV (pope)
  kidnapped by Boniface VII, 177,
    194, 195
  rejects Gerbert's appeals, 170–171
  See also Peter of Pavia
John XV (pope)
  calls for aid from Otto III, 199

  excommunicates archbishop
    Gerbert, 195–197
  supports Arnoul, challenges Hugh,
    196
  death of, 198
John XVII (pope), 217
John XVIII (pope), 217
John XIX (pope), 217. *See also*
  Romanus
John of Sacrobosco, 90–91
John XVI (antipope), 214. *See also*
  Philagathos, John
John Paul II (pope), on Gerbert, 248
John the Baptist, heads of, 237
Joseph the Wise of Spain, 56, 60–61
Josephus, Flavius, 23, 74

Al-Khwarizmi, Muhammad ibn Musa
  calculates circumference of earth,
    47, 130
  The Compendious Book on
    Calculation, 46
  as influential mathematician,
    46–48
  On Indian Calculation, 46, 89, 90
  and terms algebra, algorithm, 46
  uses, writes about, astrolabe, 57–58,
    133–134, 136
  Zij al-Sindhind, or star tables, 47,
    134, 136

Lactantius, 125, 129, 130–131
Lady Geometry, 117–118, 126
Laon, France, 186–187, 190
Last Emperor legend
  Otto III's role, 225–226
  symbolized by processional cross,
    200 (fig.)
Lateran palace in Rome, 203, 229,
    232, 235, 242, 244
Latin
  classics, 24, 36–39

as written, spoken, language of learning, 4, 32, 33–35

Law of Goths, 42–43, 44

Leo of Synada (archbishop), 194, 214–215

Leofsin, 80, 81, 117

Leonard of Pisa. *See* Fibonacci

Letaudus of Micy, 149

Libellarii (little books, written contracts), 160, 220

Liberal arts. *See* Seven liberal arts

Libraries
  of monasteries/cathedrals, 23, 49, 55, 57–58, 60, 147, 159, 165
  in Cordoba, 4, 49–50
  national library in Paris, 31, 60
  See also Scriptoria

Liege cathedral school, 111, 139

The Life and Voyages of Christopher Columbus (Irving), 129–130

The Life of Bernward (Thangmar), 32, 228–229

The Life of Gauzlin, 149

Life of Heraclius, 123

The Life of King Robert (Helgaudus), 68

Life of Meinwerk, 156

The Life of Saint Gerald of Aurillac (Odo), 13, 24

The Life of Saint Odo (John of Salerno), 24

Life of Saints Eucharius, Valerius, and Maternus (Remi), 117

Lives of the Popes abridged (Abbo), 97

Lobet of Barcelona (deacon), 58, 144–145, 147

Lorraine, duchy of, 172–174

Lothar (king of France)
  considers Adalbero, Gerbert, as spies, traitors, 66–67
  conspired against by Adalbero, Gerbert, 179, 182–183
  joins coalition against Henry, 172–174
  rivalry with Hugh for kingship, 179–184
  death of, 184

Louis V, "Louis Do-Nothing" (king of France)
  attacks traitors Adalbero, Gerbert, 66, 185
  replaces father Lothar as king, 142, 184
  as wastrel with failed marriage, 181–182

Lupus of Ferrieres, 36

Luther, Martin, 244

Macrobius, 32, 104, 108, 127

Magdeburg in Germany, 153, 204, 208

Magician Pope, 243, 245–247

Magyars of Hungary, 6, 219

Majesties
  of Saint Foy, 21, 22, 53, Plate 1 (center section)
  of Saint Gerald, 21–22, 23, 243

Malarial fever in Rome, 202, 208, 216, 230

Al-Mamun, 46, 47

Manlius, M., 31

Al-Mansur the Victorious
  sacks Barcelona and Girona, 142, 223
  regency destroyed by Berbers, 239

Map, Walter. *See* Walter Map

Mapping
  calculated by al-Khwarizmi, 47
  with earth shown as sphere, 126–128, Plate 7 (center section)
  of spheres onto flat surfaces for astrolabes, 136–139

Marozia of Rome, 194
Martianus Capella, 24, 86, 108, 117, 126
The Marriage of Mercury and Philology, 24
Martin the Pole, 243–244
Maslama of Madrid, 59, 134
Mathematical Association of America, 248
Mathematics
    advances in trigonometry by al-Buzjani, 89
    calculus in Archimedes codex, 111–112
    as form of worship, 3–4, 100
    and Indian (Arabic) numerals, 5, 46, 58–59, 79, 80, 82, 85 (fig.), 88–91
    as quest for unity given by God, 101–103, 106
    terms algebra, algorithm, 46
    today's math influenced by Gerbert, 90, 248
Mathesis
    Gerbert's mastery, 64, 75, 195
    as term, 59, 145
Matilda (princess)
    abbess of Quedlinburg, 216
    daughter of Otto II and Theophanu, 155
Memorization
    as key to speaking, writing, 36
    mnemonic poems, 80, 85, 88, 91, 143
    of psalms cycle, 33
    required for finger counting, 86
    techniques, 37, 84
    of Victorius's times table, 94
Method (Archimedes), 111–112
Mettlach monastery, 80–81, 117
Michael Scot, 134–135
Michelet, Jules, 246, 249

MicroCAD International Science Conference, 247–248
Micy monastery, 148–149
Middle Ages or medieval world
    with astrolabe as most popular instrument, 47
    as term, 129
    changes linked to fables created about Gerbert, 240–244
    See also Dark Ages
The Minor Arts (Donatus), 35
Miracles
    occur at relic jamborees, 237–238
    with opening of Charlemagne's tomb, 227
    of Saint Foy, 20–22
    of saints' relics, 149
    through Adalbert, 226
    worked by Gerald the Good, 12–13, 41
Miro Bonfill (bishop)
    and Garin, 56, 57
    profiled, 59–62
    writes astrology treatises, 145
    writes Ripoll cathedral consecration speech, 58
    death of, 171
Monasteries and cathedrals
    cathedral rules compared to monasteries, 67
    with guest houses for travelers, 18
    music as Gregorian chant, 105
    punishments for infractions, 17, 33
    See also Benedictines
Monastery schools
    curricula, procedures, 5, 33–37, 39–40, 73, 90, 101, 247
    nine-number system used to teach math, 5, 90–91
    puzzles, story problems, 202
    shift from abacus board to algorithm, 90

Monastic reform movement
  with Abbo as protector of monks'
    rights, 96, 192
  Garin's operations, 56
  led by Odo of Cluny, 24, 56
  linked to reading, book production,
    24
  ultimately transforms Church
    structure, 217
Muhammad ibn Musa al-Khwarizmi.
    See Al-Khwarizmi, Muhammad
    ibn Musa
Music
  with *Carmen Figuratum*, 165–166
  monochord as teaching tool, 73,
    104–106
  with numbers and sacred, mystical,
    104–105
  as one of quadrivium disciplines, 39
  theory of Boethius, 70, 84
Music of the Spheres (*musica
    mundana*), 104
Muslims
  as local guides through Alps, 64
  as People of the Book, 50
  as rulers of Spain, 43–44, 48–52
  ruler destroys Holy Sepulchre,
    238–239
  See also Al-Andalus, Islamic Spain

Natural History (Pliny), 25
Necromancy. *See* Black magic
Nectanabo (wizard), 146
Nicomachus of Gerasa, 101
Nilus the Hermit (ascetic), 204,
    214–216
Nine-number system. *See* Arabic
    numerals (nine-number system)
Nithard of Mettlach (abbot), 80–81
Notger of Liege (bishop), 103, 109,
    172–173
Nouaille monastery, 149

Number theory, 101
Numbers
  as keys to wisdom, power, 62, 212
  as mystical with spiritual properties,
    100–102
  See also Arabic numerals; Finger
    counting or finger numbers;
    Indian numerals; Roman
    numerals

Obertenghi family, 160
Odilo of Cluny (abbot), 162, 196
Odo (king of France), 168, 179
Odo of Cluny (abbot), 13, 63
  as reader of books, 23–24
  as religious reformer, 24, 56
Oliba of Ripoll (abbot), 57–58, 61
On Arithmetic (Boethius), 70,
    101–103, 108, 207, 212
On Astrology (Boethius), 30, 159
On Astrology (Manlius), 31
On Indian Calculation (al-Khwarizmi),
    46, 89, 90
On Medicine (Dioscorides), 243
On Music (Boethius), 70, 104–105
On Reason and the Uses of Reason
    (Gerbert), 208, 212
On Rhetoric (Victorius), 31
On the Art of Speaking (Cicero), 31
On the Diseases of the Eye
    (Demosthenes), 23, 31
*On the Multiplication and Division of
    Numbers* (Joseph the Wise), 56,
    60
On the Nature of Things (Isidore), 124,
    127
On the Reckoning of Time (Bede), 86,
    99
On Weights and Measures (Priscian), 31
Orbits of planets, sun, and moon, 5,
    121–123, 168, Plate 6 (center
    section)

Organ pipes, physics of, 5, 103–104, 106, 148

Orosius, 23, 51, 243

Otric (schoolmaster of Magdeburg), 153, 156–158, 204

Otto I, the Great (emperor)
engages Gerbert as tutor for Otto II, 6, 64–65
marries Adelaide, rules Italy, 162
and popes John XII, John XIII, 194–195
and Theophanu, 155

Otto II (emperor)
appoints Gerbert abbot of Bobbio, 6, 158
defeated by Saracens, captured by Greeks, 162–164
ivory carving, 180 (fig.)
marries Theophanu, 65
as new Constantine or Charlemagne, 211
stages debate with Otric, Gerbert, 156–158
death of, 169

Otto III (emperor)
background, profile, 199–202
battles Slavs, 209, 211
as child, taken hostage, 164–165, 171–174
coronation, 201–202, 224, Plate 8 (center section)
friendship with Bernward, 221
with Gerbert as teacher, friend, secretary, counselor, 6, 201, 210–211, 213
opens Charlemagne's tomb, 226–228
and Philagathos's mutilation, 215–216
role as Last Emperor, 225–226
supports John XV over Gerbert, 196

death of, 218, 230
successors battle for precedence, 236–237

Palimpsest of Archimedes codex, 112

Papacy
as Antichrist in Arnulf's tirade, 193–194
controlled by Marozia, 194
dangers of, 213
Gregory VII acquires power, 241
as political, not religious, position, 194, 217
supreme powers vs. equality with bishops, 193–194, 196–198, 217, 221–222

Papal bulls, 26, 64, 75
forged by Abbo, 207

Paper making, 49

Papyrus, 25–26

Parchment
production described, 25–28
vs. papyrus for book-making, 25–26

Paris, France
becomes chief city with Hugh Capet's inauguration, 186
national library, 31, 60
sacked by Otto II, 180

Peace of God, 237–238

Penance and forms of compensation for sin, 18

Peter of Pavia (pope), 160–161, 194–195, 245. *See also* John XIV

Petrarch, 129

Petroald of Bobbio (abbot), 160, 170, 222

Philagathos, John (antipope)
captured, tortured, by German partisans, 214–216
replaces Gregory V, 205, 206, 213–214
See also John XVI

Philosophy, defined, 157–158
Phisicis (meteorologists or astrologers), 116, 144–145, 146
Physics as subdiscipline of mathematics, 123, 157–158
  of organ pipes, 5, 103–104, 106, 148
Pipe organs, 5, 106, 165–166, 212, 249
Place-value system of arithmetic introduced by Gerbert's abacus, 82, 87
  originates in India, 46
  as radically new, 84, 94
  transitions to algorithm, 90
Planisphere (Ptolemy), 59, 134, 137
Plato, 101, 117, 125, 157
Pliny, 25, 30, 123, 125
Poetry
  on Boethius by Gerbert, 80
  on Cordoba by Hrosvit, 50
  on logic by Fulbert, 37
  mnemonic, on names of numerals, 80, 85, 88, 91
  mnemonic, on scientific concepts, 143
  puzzle composed for Otto II, 164 (fig.), 165–169
  The Song of Roland, 43–44
  used for instruction in schools, 36
Poland, 199, 226
Polish Catholic Church, 6–7, 219
Popular Algorithm (John of Sacrobosco), 90–91
Porphyry, 23, 38
Prince of Rome. *See* Crescentius of the Marble Horse
Priscian, 31, 35
Protestants, 244–245
Prussia, 7, 205, 208–209, 219
Ptolemy
  and celestial globes, 117 (fig.), 133

mapping method used by al-Khwarizmi, 47
  Planisphere, 59
  uses, explains, astrolabe, 131, 134, 137
Puns, 60–62 (fig.), 145, 241
Puzzles and story problems
  Carmen Figuratum (Gerbert), 164 (fig.), 165–169
  squaring the circle, 111–112
  Stomachion (Bellyache), 112
  as teaching techniques, 93–94
Pythagoras, 101–102, 123, 125

Quadrivium (mathematical disciplines), 39–40
  arithmetic and computus, 93–94, 98–99
  Gerbert's teaching at Reims, 73, 79–81, 90, 113, 120, 122
  with monochord as visual aid, 104, 105
  textbooks, 94, 108
  use of abacus, astrolabe, celestial spheres, 90, 113, 120, 122, 143
  See also Arabic numerals, Astronomy, Geometry, Mathematics, Music
Quedlinburg abbey, 216, 223, 226

Racemundo (bishop), 50, 51, 243
Ragimbold of Koln, 111, 139, 143, 147
Al-Rahman, Abd. *See* Abd al-Rahman
Rainard of Bobbio, 170
Ralph of Laon, 85, 90
Ralph the Bald (Rodolfus Glaber)
  describes eclipse, 123–124
  records signs of Apocalypse, 224, 240, 245
  on the round shape of earth, 126
Ramsey abbey, 95

Al-Rashid, Harun, 45–46

Ravenna
  Basilica of San Vitale, 154–155
  with Gerbert as archbishop, 1–3,
    216

Raymond of Laval (schoolmaster of
    Aurillac)
  correspondence from Gerbert, 38,
    174, 188
  as Gerbert's teacher, 32, 39, 57

Reconquista (Reconquest), 239–240

Reichenau monastery, 23, 30

Reims cathedral
  Abbo studies under Gerbert, 96
  Adalbero's renovations, 66–67
  archbishopric contested between
    Arnoul, Gerbert, 188–190, 191,
    195
  Arnoul confirmed as archbishop,
    222
  astronomy pursued, 113
  attacked by Louis V, 185
  becomes pawn in battle for France,
    178
  Gerbert serves as Theophanu's spy,
    171, 174–175, 178
  library, 165
  as proto-university, 68
  school, 6, 72, 73–76, 153

Remi of Trier, 115–117, 150,
    187–188

Remigius of Auxerre, 24

Renaissance rediscoveries of math and
    science, 7

Republic (Cicero), 117, 127

The Republic (Plato), 101

Rhetoric
  Gerbert's debate on reason, 208
  Gerbert's table, teaching methods,
    71, 73
  and Gerbert's writing style, 37
  as important to affairs of state, 69

Richer of Saint-Remy
  on Arnoul's confession of guilt, 193
  on Ato as mathematician, 45
  on celestial globes, 114–115
  on Gerbert as teacher, intellectual,
    59, 64, 73–74, 105–106, 240
  on Gerbert's abacus, 52
  as inaccurate historian, 74–75
  on Otric–Gerbert debate, 156–157
  relates river crossing incident, 63–64
  reports on Hugh's acquiring throne,
    186

Ripoll 225 manuscript, 57–58, 59,
    143–144

Ripoll monastery, 57–59

Robert the Pious (king of France)
  described, 205–206
  excommunicated over bigamy,
    incest, 206–207, 217
  succeeds King Hugh Capet, 205
  tutored by Gerbert, 68–69, 146,
    186
  condemns churchmen to burn at the
    stake, 238

Rodolf of Liege, 111, 139, 143, 147

Rogatus (Gerbert), 103–104. *See also*
    Organ pipes

Roman Catholic Church
  denounces enlightened monks as
    heretics, 238
  faces challenges to Dark Legend,
    244–246
  and Middle Ages period, 129
  protestants use Gerbert for
    antipapalism, 244–245
  on round shape of earth, 124, 126
  schism with Eastern Orthodox
    churches, 7, 240
  intolerance rules, issues call to arms,
    238
  mathematics as worship, 3
  not anti-science, 3, 102

relics, not education, become focus, 237

Roman Empire. *See* Holy Roman Empire

Roman numerals, 87–88, 99

Romance dialects, 34–35

Romans
choose Pope Benedict, 194
and death of John XIV, 177
rebel against Otto III, Gerbert as pope, 228–229

Romanus (pope), 217. *See also* John XIX

Rome
attacked by Slavs, 208
described, with ruins, disease, 194, 202–204, 205
Otto III seeks to conquer, falls to Crescentius III, 215, 228–230, 232
sacked by Visigoths, 42
and travels of Borrell, Ato, Gerbert, 64, 65, 75

Romuald (ascetic), 229–230

Rozala of Italy (princess), 206

Rule of Saint Benedict
described, 15–17
emphasizes book-making, reading of books, 23, 24, 25–30
includes care for travelers, 18
infractions, punishments, 17
prescribed prayers, 113
reinstated by Odo of Cluny, 24
requiring work, 25

Rule of silence, 16–17

Saint Augustine
on Christians talking nonsense, 124–126
dismisses Antipodes theory, 128, 130
explanations used in geometry text, 108

on knowledge and understanding, 233–234
warns against making predictions, 1–2

Saint Benedict, relics of, 96

Saint Foy, 20–22, Plate 1 (center section)

Saint Gerald of Aurillac, 166, 246. *See also* Gerald the Good

Saint James the Apostle (Santiago), 19, 239

Saint Luke, in the evangelary of Bernward of Hildesheim, 34 (fig.)

Saint-Columban monastery, Bobbio, 158–160

Saint-Gall monastery, 156

Saint-Gerald's monastery and cathedral, Aurillac
book-making technology described, 25–26
Gerbert's education, 15–18
new church constructed, 22–23
as stopping place for pilgrims, 19–20

Saint-Michael-of-the-Needle church, 19 (fig.)

Saint-Peter's basilica, Rome, 194, 201, 203, 247

Saint-Remy monastery, Reims, 74

Sallust, 32, 74, 142

Saltus Gerberti (Gerbert's Leap), 103, 106

Santa-Maria cathedral, Cosmedin, Rome, 22

Sant'Apollinaire monastery, Classe, 229

Saracens
associated with abacus, 52
attack Borrell in Barcelona, 186
defeat Otto II in Italy, 162, 165
wizard fable created by William, 242

Sasbach estate, Germany, 209–210

The Schism. *See* Great Schism of 1054
Science
  in Arabic language, 4–5, 45–47, 51,
    57–58, 89, 136, 248
  from Baghdad, Islamic Spain, to the
    north, 5, 45–46, 59, 76
  books in Ripoll library, 57–58
  of Islam passed to Christian West,
    140, 248
  paired with faith dies with Otto III,
    102, 231
  with physics as separate discipline
    from math, 123, 157
  rediscovered during Renaissance, 7
  as requirement for popes, 194
  See also Education, Quadrivium
Science-Religion war, 131
Scientist Pope, 3, 218–219, 237, 246
Scriptoria
  described, 15, 25
  in Echternach monastery, 80
  and Petrarch's discoveries, 129
  in Ripoll monastery, 57
  See also Libraries
Searcher magazine, 248
Sedulius the Scot, 35
Seniofred "Lobet." *See* Lobet of
  Barcelona
Sergius IV, "Peter Pig's Snout" (pope),
  217, 235 (fig.), 236, 240
Seven liberal arts
  described, 39
  learned and taught by Gerbert, 68,
    240
  Martianus Capella's textbook, 24,
    117
  at Reims cathedral school, 65
  as seven streams of wisdom, 50
Severus Sebokt, 88
Siege engines, 213, 215, 229
Sigebert of Gembloux (abbot),
  240–241

Sign language used by monks, 16–17,
    87. *See also* Finger counting or
    finger numbers
Sky & Telescope magazine, 248
The Song of Roland epic poem,
    43–44
Song of the Algorithm (Villa Dei),
    90–91
Sophie (princess)
  daughter of Otto II and Theophanu,
    155
  and dispute over Gandersheim,
    220–222
Spain
  and Gerbert's papal acts, 223
  invaded by Muslims from Maghrib,
    43
  peace destroyed by Muslim-
    Christian fighting, 239
  See also Al-Andalus, Islamic Spain
Spheres/celestial globes
  armillary, 5, 121–123, 136, 138
  compared to astrolabes, 133–134,
    139
  for constellations, planetary motion,
    73–74, 119–121
  constructing, 114–115, 118–119
  history and description, 117–120,
    117 (fig.)
  for Remi of Trier, 115–117, 150,
    187
  used to keep time, 113–114
Stabilis of Orleans (Constantine of
    Fleury), 70, 147–148
Star Tables or *Zij al-Sindhind* (al-
    Khwarizmi), 39, 47, 134
Stephen (king of Hungary), 249
Al-Sufi, 118
Sundials, 114, 136
Sweden, 7, 219
Sylvester II (pope)
  appointed by Otto III, 218

finally credited with science, math,
advances, 247–249
papal acts, 219–224
black magic myths, 232–235
dies after Otto III's death, 218,
234–236
See also Gerbert of Aurillac
Synesius of Cyrene (bishop), 136–137

Thabit ibn Qurra, 46
Thales of Miletus, 124
Thangmar of Hildesheim
(schoolmaster), 32, 221
The Jewish War (Josephus), 74
Theodoric (king of the Ostrogoths), 38
Theophanu (empress)
with Adalbero, Gerbert, against
Lothar, 178, 182–183
animosity toward Adelaide, 196
background and profile, 154–156
devises rescue of Otto II from
Greeks, 163–164
fails to support Gerbert, 189
has Gerbert spy from Reims,
174–175, 178
marries Otto II, 65
with Philagathos as companion,
213–214
portrayed in *Carmen Figuratum*, 167
(fig.)
as regent for Otto III, 172–173,
174, 180 (fig.)
travels with Otto II and Gerbert,
154–155
Thietmar of Merseburg (bishop)
on Gerbert's knowledge, 240
on Otto II, 155–156, 162–164, 171
on Otto III, 226
refers to *horologium*, 135–136, 208
Timaeus (Plato), 117
Time measurements, 113–114, 136,
138–139

Tolerance creed (*dhimma*) in
al-Andalus
for Christians, Jews, Muslims, 50,
51
Trajan's column, 202, 221
Travels
in al-Andalus, 48
from Aurillac to Vic, 53–54
of Gerald the Good, 64
hazards, tolls, crimes, 63–64
of Last Emperor to Jerusalem, 200
monastery guest houses, 18
of Odo of Cluny, 24, 63
of Otto III, Gerbert, to Rome,
198–199, 220
pilgrimages to Compostela, 19–22
of Sophie, 220
of Theophanu, Otto II, Gerbert,
154–155
Treason accusations
charged against Boethius, 38
directed at Adalbero, 142, 183
directed at Gerbert, 6, 142, 183
Triangles
Boethius on Pythagoras, 101
Gerbert calculates area, 3, 5, 46,
109–111
interior, exterior angles, 111
as mother of all figures, 109
See also Geometry
Trier monastery and cathedral,
116–117, 239
Trigonometry, 89
Trivium disciplines, 39, 73
Tryggvason, Olaf (king of Norway),
219

Ummayad clan, 43
Unity in mathematical order, 100–102,
106, 108
Universe, models of, 5, 121–123, 168,
Plate 6 (center section)

Vajk (king of Hungary), 219. *See also*
    Stephen (king of Hungary)
Venerable Bede, 23
    commentaries by Abbo, 97
    earth as egg-shaped, 126
    on finger counting, 86–87
    incorporates A.D. dating system, 99
    on round shape of earth, 126
Verdun, 182–183, 186
Vic, Spain, 45, 54–55
Victorius of Aquitaine, 31, 94, 99, 100
Vigier, Geraud, 246
Vigila, 58–59, 89. *See also* Codex
    Vigilanus
Vignier, Nicolas, 244–245
Vikings
    attack Otto III, 199
    Christianized by Otto III, Gerbert, 6
    forms of compensation for crimes,
        18
    sack city of Saintes, 21
Villa Dei, Alexander de, 90–91
Virgil
    complete works in parchment, 26
    Gerbert's copy, 36
    serpents of, 24
Virgin Mary, 1, 202
Visigoths
    code of law, 42–43, 44
    encouraged to settle Catalonia, 44
    and Isidore of Seville, 42–43, 124
    kingdom of Spain, 43–44
Vladimir (prince of Kiev), 219

Walter Map, 241–242
Wazzalcora (astrolabe), 133, 143
Whewell, William, 131
White, Andrew Dickson, 131
Widukind, 161–162
Wilberforce, Samuel (bishop), 131
William I of Aquitaine (duke),
    "William the Pious," 12
William of Arles (count), 182
William of Malmesbury
    on abacus and astrolabe of Gerbert,
        52, 89–90, 134, 145
    on Gerbert's death, 234–235
    writes wizardry fables about Gerbert,
        145, 232–235, 242–243
Willigis of Mainz (archbishop)
    in coalition against Henry the
        Quarreler, 174
    and Gandersheim convent conflict,
        220–222
    as guardian of Otto III, 165,
        171–172
    proclaims Henry of Bavaria as king,
        231–232
    as Theophanu's adviser, 189
Wizardry. *See* Black magic
World Year of Astronomy 2009,
    248–249
Writing taught in monasteries, 33–34

Zero as place holder, 85, 90
Zij al-Sindhind (al-Khwarizmi), 39
Zoe of Constantinople (princess), 231